In memory of Carol who inspired me to learn and grow.

Critique

of

Pure Experience

by Richard Avenarius

VOLUME 1

translated by David Grunwald

2018

Original Third Edition

Leipzig 1921

Foreword to the Third Edition of the First Edition

The world of thought, its mechanistic nature becoming a clearer and stronger form of mechanical materialism most recently effecting the old doctrines, posits that all happenings ultimately occur from pressure and shock based on attraction and repulsion. There is no question that the mechanical theories, with their imposing foundation of absolute space, time and movement especially those of Copernicus, Kepler, Galileo and Newton, but also by Descartes and Spinoza, achieved greatness under their leadership in the eighteenth and nineteenth centuries.

Ever higher levels of development in physical science, technology and industry were achieved under their leadership of the economic doctrine of Adam Smith which forged a harmony of industrial and commercial interactions and influence: namely the free competition of commerce and a sociological image of competing forces of atoms. From the economic side of anthropology, they subjugated the entire organism hierarchy with Darwinian theory that stated a species had survived had adapted well to their environmental conditions which created an analogous competition for food. Yes, even psychology conquered itself in the doctrine of the association of ideas. No question, then that they had a profound effect and brought substantial enlightenment.

But they wanted greater things than they had in their power. To be sure, increasingly they revealed the mechanical side of reality, but they took these for the most important things and closed their eyes to those problems for which they had no answers.

Let us see from the deepest philosophical insights, which sadly did not receive enough attention, this then was the earliest field of Sociology: the growing dissatisfaction and the need for increased employment of the working masses which made a mockery of the harmony of economic interests and was indeed the last reason for the world war. Then it moved to Biology: Vitalism — a reaction, not understood by its representatives as such

— to the inadequacy of the mechanical view of nature; it is not Solution, but only a symptom of a difficult situation. [1] The solution must be taken from the source of mechanistic errors. For insight into the latter, philosophy has already won a hundred years. Meanwhile it was the authority of Newton's mechanics that prevented it from becoming fertile. The contrast of the primary and the secondary qualities were — if only in favor of the latter, rather than for the sake of obtaining an indifferent neutral intuition — from Berkeley[2] and also already overcome by Leibniz[3]. Mach[4] put himself on the same ground fifty years ago and his "mechanics" became more and more the epistemological basis of a "phenomenological" physics. However, the right time for the effectiveness of this most powerful transformation of human thought only began as the mechanical "view of nature" even physics, faced the most embarrassing contradiction: before those of Fizeaü and Michelson and, when Einstein took the theory of relativity, Mach at the end of his own monumental doctrine, came to terms that was no longer compatible with the primary qualities of things. Now we are in the midst of a change, and the hour of the mechanical view of nature has struck. The general cultural meaning of the theory of relativity is that it applies physics to the ground prepared by Berkeley, their coincidences of perceptions to the last basis of experience and thus approaches the physiology of the senses and biology in general. This begins to close the almost culturally hostile gap between the physical and biological sciences. And thus, we recognize more and more the high cultural heritage we have inherited in the life work of Richard Avenarius. As the theory of relativity lays a bridge from physics to biology, so does Avenarius' pure theories of vitality from psychology and the epistemology of biology to the social sciences and humanities. This offers the prospect of gradually overcoming the fracture of knowledge into the common and the unified, not merely conceptual, but also a closer living reality that can thereby be made useful.

Notes:

[1] Vitalism is the belief that "living organisms are fundamentally different from non-living entities because they contain some non-physical element or are governed by different principles than are inanimate things. Where vitalism explicitly invokes a vital principle, that element is often referred to as the "vital spark", "energy" or "élan vital", which some equate with the soul. In the 18th and 19th century vitalism was discussed among biologists, between those who felt that the known mechanics of physics would eventually explain the difference between life and non-life and vitalists who argued that life could not be reduced to a mechanistic process. Some vitalist biologists proposed testable hypotheses meant to show inadequacies with mechanistic explanations, but these experiments failed to provide support for vitalism. Biologists now consider vitalism to have been refuted by empirical evidence, and hence as belonging to the realm of religion rather than that of science.

[2] George Berkeley (12 March 1685 – 14 January 1753) — known as Bishop Berkeley, was an Irish philosopher whose primary achievement was the advancement of a theory he called "immaterialism" or later known as "subjective realism". This theory denies the existence of material substance and instead contends that familiar objects like tables and chairs are only ideas in the minds of perceivers, and, as a result, cannot exist without being perceived.

[3] Gottfried Wilhelm (von) Leibniz (1 July 1646 – 14 November 1716) was a German polymath and philosopher who occupies a prominent place in the history of mathematics and the history of philosophy, having developed differential and integral calculus independently of Isaac Newton. Leibniz's notation has been widely used ever since it was published. It was only in the 20th century that his Law of Continuity and Transcendental Law of Homogeneity found mathematical implementation (by means of non-standard analysis). He became one of the most prolific inventors in the field of mechanical calculators. While working on adding automatic multiplication and division to Pascal's calculator, he was the first to describe a pinwheel calculator in 1685 and invented the Leibniz wheel, used in the arithmometer, the first mass-produced mechanical calculator. He also refined the binary number system, which is the foundation of virtually all digital computers.

[4] Ernst Waldfried Josef Wenzel Mach (18 February 1838 – 19 February 1916) was an Austrian physicist and philosopher, noted for his contributions to physics such as study of shock waves. The ratio of one's speed to that of sound is named the Mach number in his honor. As a philosopher of science, he was a major influence on logical positivism and American pragmatism. Through his criticism of Newton's theories of space and time, he foreshadowed Einstein's theory of relativity.

Author's Forward by Dr. Richard Avenarius

In my Spinoza paper I endeavored to regard the development of a particular worldview as a legitimate process from a purely psychological point of view; in the forwarding remarks of the present work there was an attempt to see the source, task, method and design of an entire philosophy as determined by a general principle. The work finally attempts to conceive of all theoretical behavior in general – in itself and in its relation to the practical, as well as in the general — as a consequence of a single simple presupposition.

On such an extended basis, the experience to which the inquiry should always remain directed was only to be treated as a special case, which, to be sure, seemed to require that of corresponding "general" considerations. — The meaning of experience as a special case received special treatment during the entire investigation.

Another consequence of the suggested approach was that new formal contexts of cognition were revealed and more new material was introduced, thus the more the interest in the inner unity of all human activity developed — however, the interest in the questions relating to experience, how these are formulated by schools of thought, and unfortunately, their association with basic concepts act as a sanction to tradition. I make apology for referring to the work itself where such shifting interests offer a more general "explanation".

In any case, I now have to be prepared that I will not be spared the accusation that the critique of pure experience has frighteningly little to do with the serious and important questions, such as those of other philosophical investigations of object-related nature. In fact, it is simple problems that this criticism raises, so simple that the "true critical philosopher" will look down on them with compassionate pride. But at least it should be problems that, even if they value dignity and worth far behind the others, at least as regards the time spent on their treatment, they wish to gain the upper hand, inasmuch as they appear concerned with simpler and more general inquiries.

First a few words about the justification for the whole experiment!

It may not be impractical to immediately send two preconditions which I, for the moment and without stressing this term, would like to name empiriocritical axioms: the first is the axiom of knowledge-content, the second the axiom of forms of knowledge.

These two conditions could be formulated something like this:

1) Every human individual initially assumes an environment of diverse components, along with other human individuals with varied statements, and what is expressed as dependence on the environment: all the cognitive contents of the philosophical worldviews – whether critical or not — are **modifications** of that original assumption.

With the latter it can be said: this resulted in e.g., a Plato, a Spinoza, a Kant — the philosophers won their results by positive or negative augmentations of that assumption which they too made at the beginning of their development.

2) Scientific knowledge has no substantially different forms or derivations than the non-scientific: all special forms of scientific knowledge or means derive from pre-scientific training.

With the latter is can be said: methods brought from mathematicians and mechanics — in the final analysis must be reduced to simple and generally human functions.

For the readers who agree with these two sentences — I hope the goal of this project — and this is the only reason — will motivate them faster.

Whoever agrees with the first sentence, may also admit that it is advisable to start from the original assumption in the treatment of our subject of investigation — and not later amendments to the same, e.g., not of the "consciousness" or the "thinking" as the "immediately given" or "immediate conscience" — and it may already be preparing too much for the all to ready misunderstanding that the so-called "immediate existence of consciousness" is a derivative of a theory which, as a special case as a variation of assumption, is much more diverse and perhaps also stems from very different knowledge in its historical development. From "consciousness" or "thinking" — for the purposes of the development of our own views over the recognition or even judgement of others — "the beginning" means in the best case, to avoid a more drastic comparison, starting at the end!

So much for the justification of the chosen starting point —.

Admittedly, from the original assumption it was hinted, or admitted, that one should agree that if one starts with the environment and the relationship to the individual it is again unreasonable for one to scarcely notice the "influence" or "stimuli" on the central nervous organ and immediately move from the organ changes to "consciousness" and "thinking" — to skip past the "ideas" of the individual, rather than to changes in individual "stimulus" in the agitated central organ, and then to follow the various dependent relationships amongst the organs.

In this way, I hope to answer reasonably justified questions, —
Those who then admit the second sentence will probably also be inclined to admit that it is advisable not to immediately or exclusively elaborate on complicated and special forms or reflect on sophisticated "scientific" cognition, but also on ordinary life itself with its natural and unbiased cognition from which the scientific developed, and thus we keep in mind the affinities of science with the pre-scientific forms or means.

Allowing for these possibilities, one might also be able to admit that it would be advisable to discuss the "possibility" of certain kinds of insights before attempting to describe cognition in general terms of nature and contexts: and that means that the underlying requirement is not material and special, but a formal and general theory of human cognition.

This explains why the measurements on the scale of other "critics" have so few ambitious goals in their elevated society.

And so much then for my motivation to attempt a "return" to the natural point of departure, rather than to attend to those philosophers and instead of books — I am attempting to "connect" directly with things.

I am concerned with the publication of this book and not the general justification for it. After all, it was not enough to overcome intense subjective reluctance, but also many large and objective reservations! Had not younger researchers, who became known and subscribed to the views set forth here by my lectures, urged me further and further that it became not only a *right* but a *duty* to fulfill — I admit I would have preferred it that the manifold imperfections only I could be convinced of — but God only knows for how long? — in order to promote it for the time being by constantly adding new things and making improvements in small steps in the direction of the ideal that decades ago knew its own youthful courage, for many years then I had planned for a maturing work and yet I progressed until finally retreating into the incalculable distances.

Another thing to mention was the tremendous richness of the material with which had already been struggled with. It grew incessantly — and the conversion of the old and pursuit of the new consumed more time and strength.

In order not to be crushed, I had to decide, from a certain moment onwards, to blame myself for this, or that which served life - lower and higher - or literature - older and newer – finding not enough or even to have considered some, as still more material for processing; and many things collected for some time had to be postponed. And in order to be able to solve the task as it evolved, I had to solve it for myself before finally sharing it.

This then happened in such a way, that I distinguished between a criticism and a system of pure experience and strictly separated their treatment completely.

Then I resigned myself completely from the critical examination of the views of others. What remained desirable to me in this respect would find opportunity to catch up elsewhere.

Of the many works questioned, only the "Critique of Pure Reason" is mentioned. — As I wrote the introduction to this book, I chose the title "Critique of Pure Experience" not without anticipating a polemic; today — in the service of a good bit older philosophy — I connect this book's purpose with a conscious homage to the genius of Kant. A comparison of my little work with his gigantic creation was completely remote to me before as it is even today. But of course: an exposition of the relation between the critique of pure experience and the critique of pure reason was an original intention; Now as with earlier intentions, I have returned to it. The Kantian philology of lower and higher rank has taken on a development that does not invite me — even in the particular case where I would like to take Kant's questions on directly. Namely, in which circles should I expect interest in my relationship with Kant, since I could not know if my work would interest any circles? — So, I respectfully greet the gentlemen followers of the great master Immanuael and ask them for forgiveness, if I did not even include his most powerful works in books and systems which I did not take as my task to criticize.

The decision to treat only the things and not the views about the things is then related to the fact that on the one hand I did not respond to the analysis of different, in my opinion erroneous concepts of experience. On the other hand, but also in the "*Kritik der reinen Ehrfahrung*", the decision was made to refrain from acquiring "confirmations" from other research sources. In this respect, I can say I've hardly ever pitched a book or treatise representative of scientific philosophy without enjoying any agreement on this point; but only where a suggestion or instruction was passed to me did I take it upon myself to add additional quotes. Incidentally, as the prologue shows, other authors have influenced me - if my memory is not mistaken - in the proper sense. I have only an association with W. Wundt(1)(especially to his highly useful adjustments "experiences on the functional divorce of organs with a sentence about the functional indifference of the basic components.) I was greatly encouraged by the writings of Ernst Mach(2) to which the philosophical reader should necessarily be made aware.

At last, I was too modest: not wanting to offer anything but my own personal views which may have matched other's specific observations; to offer nothing more than a tentative attempt to look at things from a different point of view, perhaps to stimulate others. *Kritik* became a crisis for me — perhaps it will also help another person to experience a fruitful crisis or help him out of a crisis that does him no good.

If I willingly reduce the value of the results which I arrived at to the importance of possible suggestions, I hope then to be able to designate the scope as less limited. At least from the beginning, I sought as an end goal the first principles of a general theory of human knowledge and to outline actions that connected science in general and in particular psychology (in the sense of psychological variations) and subsequently to prepare the soil for scientific pedagogy, logic, ethics, legal philosophy, economics, linguistics and the like. I kept in mind all these sciences when designing and executing this work; of course, one may go to or from the *"Kritik of Pure Experience"* not even expecting a science of science, psychology, logic etc., even in specially delineated areas. The suggestions of content of the general foundation of these sciences should offer their own basic concepts that are recognizable and easy to follow — at least to those who care about humanities most important assets — morality, law, science, the state as a whole, society which means the individual and the general welfare – to be put on the most certain reasoning, which after all, can only be hoped for by scientific analysis.

If it be enough for me to offer suggestions, it means that I am not claiming finished results. It would be foolish for me not to expect in the hundreds of concepts that follow, more or less large numbers of them to be conceivable in other provisions or in other classifications — and the criticism, if this book be worthy of such attention, will find here a rich field for such activity.

On the other hand, I can also hope for critical insight and further work, that will not fail to be recognized in the expositions and improvements – which are most welcome to me – the difficulties which characterize the attempt to formulate such a general theory, if only as a sketch outlining the bewildering variety of substances. One would have to prepare for great variability and ambiguity, abstractness of most concepts, lack of support for fixed points of attack and for certain methods.

For details, which are needed by some for a refresher but also to enable further education, I not only include the values dependent on changes in the nervous central organ (part II, III), but also some terms used in the first part: Here, by and by, some material expressions, which owe a temporary continuance to the convenience of their use, are replaced by other, purely formal, terms, e.g., quantitative. However, I do not consider the latter particularly urgent.

In order not to hide the fact that *Kritik* delivered final results that differed from the prologue, this will probably require no special apology. As happy as I would be if *Kritik* remained itself inconsistent, it seems to me to be irrelevant whether it contradicts the prologue in one way or another. At this point the work has merely stepped over the older work.

Regarding the presentation, an attempt was made to deal only with the matter; to take these matters as possible, and as they revealed themselves: I also counted the world views and concepts of knowledge of the people. I did everything in my power to detect and mitigate any prejudices that could affect me. I seriously tried to gain a position over the factions; to admit that everything was true did not seem more naive to me at the beginning than to have accepted that nothing was true. Holding this opinion convinced me that it was not important form me to answer old questions; instead, I asked new questions of old answers. „New" meant, at least for me, the immediate impression.

Where my object was no longer able to speak for itself, there was no rhetorical turn that helped. What I explained is openly there. I believe a careful division gives a clear overview that soon reveals mistakes in the material. If violations of logical norms creep in, there will be no hiding place even from the eye of the less critical reader.

In other respects, I have been anxious to prevent the misunderstanding of certain provisions by special technical terms. This was by no means a pleasant task; whether it was well done will be decided in the future. If one fails to maintain a relatively new concept as it is understood and wants to distinguish it from related ones by means of a new, relationship free expressions, then one must fear the familiar accusation of "new wine in old bottles"; if one forms new terms, then there is usual complaint of "impeding of understanding". In short, whatever you do, there are shortcomings. After all, I prefer not to be misunderstood; so, I opted for new terms. At the time of formulation, it was characteristic of me that, if possible, they sought to be representative — or at least useful. (For the latter reason I have not shied away even from philologically questionable forms of linguistic mutilation of the suffix and other offenses that I humbly admit will be an accusation against me). In any case, I already had the opportunity to observe how quickly, with some concessions, the proposed terms took and did their good work.

If in the first part, I renounced addressing the more specific requirements of physiological details; I knew very well what decorations I took away from my presentation; I confined myself — in the memory of Lotzes(3) — to the most general conditions of this kind, because they alone fulfilled my purposes (those of specialized research did not apply here) and seemed in the end to give a much greater guarantee of permanence than the highly specialized ones.

In order to illustrate the dependent values, in Part II, I took abundant examples of the oral or written statements of individuals in the most varied stages of development; to a more limited extent cases were admitted in which I myself entered the reader into the relationship of a statement, for example in the case of the clock (Footnotes to num. 491 vol 2)

In order to illustrate the dependent values, in Part II, I took abundant examples of the oral or written statements of individuals in the most varied stages of development; to a more limited extent cases were admitted in which I myself entered the reader into the relationship of a statement, for example in the case of the clock (Footnotes to num. 491 vol 2)

In Part I, on the other hand, perhaps some such explanatory examples will be missed. I had to be just because I had to deal with changes in quantities and series of changes in subject matter and task and remain completely unbiased in approaching this formal sphere. Also, the "psychic expression" and the "consciousness" of these changes or as one might call it otherwise, the representation of form and content, would be in a state of transformation and probably very soon be in the space where they were at risk of encountering individual differences of opinion (*individuelle*

Meinungsverschiedenheiten), subjective valuations and emotions hostile to thought. It was here, in the purely formal consideration of changes of the central nervous organ, that it seemed to me the only neutral ground upon which the questions, which are so apt to arouse our passions, could at some point be treated with complete dispassion.

Finally, when I consider how little it is, what I now present in tangible form, I feel how much more it could have been if I had kept closer to the conventional procedure. How much more — but at the same time not with the same weight! How much more, how much easier, how much more grateful! But what was achieved could only be reached on the path taken which was mine.

I was another person when I took up the staff to hike to the distant land of knowledge — and am another as I lay it down. The childish confidence I had to "find the truth" is long gone; it was only as I progressed that I experience real difficulties and limits to my powers. And the end? — — — If only I came to achieve clarity with myself!

Hottingen bei Zürich
Ostersonntag 1888.

R. A.

(1) Wilhelm Maximilian Wundt (16 August 1832 – 31 August 1920) was a German physician, physiologist, philosopher, and professor, known today as one of the founding figures of modern psychology. Wundt, who noted psychology as a science apart from philosophy and biology, was the first person ever to call himself a psychologist.

(2) Ernst Waldfried Josef Wenzel Mach (18 February 1838 – 19 February 1916) was an Austrian physicist and philosopher, noted for his contributions to physics such as study of shock waves. The ratio of one's speed to that of sound is named the Mach number in his honor.

(3) Rudolf Hermann Lotze (21 May 1817 – 1 July 1881) was a German philosopher and logician. He also had a medical degree and was well versed in biology. He argued that if the physical world is governed by mechanical laws, relations and developments in the universe could be explained as the functioning of a world mind. His medical studies were pioneering works in scientific psychology.

Forward To Dr. Richard Avenarius' Critique of Pure Experience 1888-1890
Volume I
by David Grunwald

For the first time in a century, volume 1 of "Pure Critique of Experience,"
written by Dr. Richard Avenarius is now available in English. This work,
published in 1888, forms a complete system of ideas, fragments of which can
be found in many contemporary fields such as behavioral psychology to
biometrics, computer science and psychology.

But it does more than that. In this work, Avenarius, manages to capture the
mind at play. His descriptions of the natural world are written in fluid form,
free from dogma, and formal "schools of thought". For example, he does not
reiterate works of other philosophers to support his own search. In the
work, the careful reader can rediscover the spirit in inquisition that so many
ancients exhibited. In this sense, his work rediscovers the purity of thought.
His style is flowing yet engaging. His work is unique in that there is no
hypothesis, only an investigation into the realm of pure experience.

The late Gerhard Kratzsch wrote that Avenarius originally sought to write an
Ästhetik, or book which would set out principles concerned with nature and
appreciation of beauty, especially art. Instead he gave a biological
foundation for an aesthetic to come and ideas to ponder about.

Despite this, the book isn't a quick or easy read. It conveys ideas tested in
logic that may take the patient reader several passes to understand the
concepts. But the ideas are sublime in their beauty and the logical
construction is pure – no schools of thought are employed to buttress
arguments. In constructing a theory of pure knowledge, Dr. Avenarius,
inspired by a single purpose, led the reader through the his journey. It was
his hope that this journey would itself illuminate a path toward
enlightenment at a time when many ideas and "schools of thought" began to
cloud the horizon.

Examining ideas in a new complete way isn't always clear and certainly isn't
easy. Quantum mechanics, an area prone to debate and misunderstanding is

one such example. Quantum mechanics is the mathematical framework for the development of physical theories. Like any new endeavor, establishing the postulates, or ground rules, is often long and controversial.

Nielsen and Chuang in their pioneering book "Quantum Computation and Quantum Information," capture the rocky path toward establishing axioms:

The postulates of quantum mechanics were derived after a long process of trial and error (mostly), which involved a considerable amount of guessing and fumbling by the originators of the theory. Don't be suprised if the motivation for the postulates is not always clear; even to experts the basic postulates of quantum mechanics appear suprising.[1]

Dr. Avenarius pursues defining his system C and the postulates for obtaining "pure knowledge" and in the pursuit offers us a rare glimpse of a mind at play in the classical sense.

The dawning of the 20th century, called the *fin-de-siècle* was filled with critics, notably Mauthner and Wittgenstein, each who came up with different ways to describe the usefulness of language in the acquisition of knowledge. Avenarius comes from a different angle – not pure science and math.

He wanted his knowledge pure and he tried to give us a system to do just that. He is first and foremost an *explorer*, which makes him relevant, whether we are studying quantum computing, psychology or the environment. His ideas open wide vistas drawn from a confluence of deep and rich ideas grown in the special environment of late 19th century and early 20th century firmament.

Avenarius wanted his works made available in their entirety — which means he wanted to be read. His style is unique and flowing for he was learning while teaching his discoveries. In fact, he struggled to understand concepts himself before sharing them with others.

He believed that science must be concerned with purely descriptive definitions of experience, which must be free of both metaphysics and materialism. His opposition to the materialist assertions resulted in vehement attacks by Vladimir Lenin in his work "Materialism and Empirio-criticism," a book that was required reading for many years inside the Soviet Union.

Avenarius' principal works are the famously difficult *Kritik der reinen Erfahrung* (*Critique of Pure Experience*, 1888–1890) 2 Vols and *Der menschliche Weltbegriff* (*The Human Concept of the World*, 1891) which influenced thinkers like Ernst Mach, Edmund Husserl, Ber Borochov and William James. It has been written that when Richard Avenarius, Professor of Philosophy at the University, died at Zürich on 18th August, 1896, only a very small circle of philosophers and pupils knew what a powerful mind had been snatched from amongst them; for he was a man whose unique thought was unappreciated by his contemporaries.

Avenarius constructs an entire system based on components, the environment, as it exists and a description of the systems in which humans live and struggle.

In his principal work, the "*Kritik der reinen Erfahrung*" (3ed. Leipzig 1921) – "Critic of Pure Experience", Richard Avenarius first makes the attempt to simply describe all theoretical activity, in itself and in its relation to practical activity (which he also describes more generally) as conditioned by analytically determined changes of the nervous central organ. In this way, he arrives at a formal and general theory of human knowledge and action; he endeavors to limit scientific philosophy critically to the descriptive definition of the general idea of experience in its form and context. Avenarius believed that science could be a great help in assisting in providing elemental descriptions.

The point of departure for his investigations was the "natural" or "pre-scientific assumption" of a "principal coordination" between the self and environment. Each individual finds himself facing both an environment with various component parts and other individuals who make assertions about this environment, which also express themselves. The initial principal coordination thus consists in the existence of a "central term" (the individual) and "opposite terms" about which he makes assertions. The encountering individual is represented and centralized in what he called *system C* (the central nervous system, the cerebrum), which he considered part of the basic biological processes which included nourishment and work.

In dealing with a topic of immense proportions, Avenarius was keen to reduce the scope of his investigation where possible. For example, he makes a distinction between general and special vital differences and chooses to examine only the latter. Then a further split is made of the vital differences into significant and insignificant cases. No further division was required among the significant cases. Uniformity in nutritional and work expenditure is sought to further simplify the cases selected for examination.

The result is a complete system of knowledge free from injection of preconceived notions or schools of thought.

Dr. Friedrich Carstanjen at the University of Zürich wrote in 1897, "The *Kritik der reinen Erfahrung*" is not only a theory of experience, but, inasmuch as experience is a species of the genus knowledge, it is also a theory of knowledge. And while in all special theories of knowledge philosophers endeavor to develop what in particular they mean by knowledge, Avenarius, on the contrary, aims only at presenting the common normal element in all such theories, the universal norms according to which individuals determine Being and Knowing. Ultimately, therefore, the "*Kritik der reinen Erfahrung*" is not only a general theory of knowledge; it is also a general theory of human norms."[2]

Empirio-criticism, on the other hand, takes up the position that everything is experience when it has been stated as experienced by an individual—though it may be that primarily it is only experience for this one individual in question. It has been said that for materialism to begin without a soul is a principle, for Empiro-criticism, it is a method.

Avenarius' system examines experience and knowledge from the effect of the environment (called R values) onto the individual organism (System C) which is effected by descripted objects he calls E values (e.g., hot, cold, red, hungry). The composites of fluctuating values generate statements e.g., "The sunshine makes me feel happy."

E-values contained in Avenarius' System C reduce the unknowns to knowns. It is in the descriptiveness of these changing components that he finds a use for science as a tool. E-values are constantly changing and acquiring new values from environmental values (R-values); however, changes in R-values do not necessitate changes in E-values in every case.

For example, if an individual says, "I have the perception blue," it then may be approached from the side of the designation or from the side of that which is designated. The color and perception can be examined as separate or in relation to changes in the environment or sub-systems he describes in detail in his book.

Beginning his analysis with the end state, he accepts no preconditions or presuppositions only the pure state existing at *a given point in time*. In the system, he recognizes that oscillations (Schwankung) occur continuously and can be viewed as ordered (1st, 2nd, 3rd, etc..) vital series.

Adjustments of System C in annulling its changes may be most varied in kind. In the first volume of the "Kritik den reinen Ehrfahrung" Avenarius submits them to a searching investigation. Changes conceivable in System C are divided into ectosystematic and endosystematic. Changes are called ectosystematic which, though their first phases occur in System C, complete their course outside of it, as in movements of the limbs. Those changes are called endosystematic which take place entirely within System C. When, for instance, something is lost, the vague running and searching for it depends upon ectosystematic changes; the reflective consideration of the circumstances in which it was mislaid or lost depends upon endosystematic changes. When philosophers try to solve the question as to the "origin of consciousness" by "thought," this solution depends upon endosystematic changes; when the physiologist, on the other hand, institutes practical experiments in reference to this problem, it is by means of ectosystematic changes that the vital-difference is annulled.

The organism labelled as "System C" itself makes adjustments in several ways. (1) It reduces unknowns to knowns by a method of super ceding one set of understandings for another. Humans will typically reduce to the highest general concept to settle to a "known". (2) The concept of gradual habituation results in a process as unknowns are sifted to knowns. The resting point may itself reveal new unknowns requiring the processes of reduction to knowns once more. (3) Temporary changes may be adopted in order to facilitate the process of changing unknowns to knowns.

E-values forming to statements showcase dependencies as to their strength, pitch and timbre. Fluctuations or oscillations are constant as changes necessitate compensating movements which complete nullification of any given vital process seeking an equilibrium, a balance, thus accomplishing a type of "systemic preservation". A statement of strength depends on the amplitude of the oscillation. A statement's pitch can be ascertained by the number of oscillations and the statement's timbre depends on the form of the oscillation. For Avinarius, the change process in the System C organism falls into 5 categories:

1. General sense form. These include sensory e-values like touch or feel.
2. Magnitude and intensity. These include strength or measure of divergence from equilibrium.
3. Direction of change of the process.
4. Relevancy of the stimuli which imposes a significance as to the effect of the part on the whole.
5. Familiarity influenced by habit which resolves to "real", "known", or "certain" for example.

Avenarius presents a complete view of the universe as one, which corresponds to theoretical as well as to practical needs. He comprehends all our action and thought as E-values, which depend immediately upon the change-processes-preservation in System C, and immediately upon the change-processes in the component parts of the environment (R-values).

In finishing this translation of the first volume, many hours were spent getting to know Richard Avenarius, his style and his thoughts. The effort to express his *Heiterkeit*, or essence was a conscious decision throught the translation. His proclivity to play and learn by trial was inspirational. Rereading the words provided an awareness of thoughts and concepts I treasure and hope that the reader finds new insights from old wine in new bottles!

To me, the effort brought out another dimension, this one from an unlikely person in Percy Shelley, the English poet. I think Avenarius would have agreed with the poet Shelley's observations:

"What is life? Thoughts and feelings arise, with or without our will, and we employ words to express them. We are born, and our birth is unremembered and our infancy remembered but in fragments. We live on, and in living we lose the apprehension of life. How vain is it to think that words can penetrate the mystery of our being. Rightly used they may make evident our ignorance of ourselves, and this is much. For what are we?"

I hope you enjoy the journey of discovering the ideas of Richard Avenarius. You will learn something new every time.

David Grunwald
Campbell, California 2018

[1] Nielsen, Michael A., Isaac L. Chuang "Quantum Computation and Quantum Information", Cambridge University Press, 2016

[2] Carstanjen, Friedrich Dr. "Richard Avenarius and his General Theory of Knowledge Empiriocriticism," Zürich 1897 translated by H. Bosanquet 1906

Critique of Pure Experience

Contents

___

INTRODUCTION

GENERAL TASKS OF THE CRITQUE
OF PURE EXPERIENCE

I.

1. — The object, task, character, method, position and prerequisites of our investigation presently for the time being – we note the relevance of individuals no less than meaningful historical declarations – contain the assumption:

it is an arbitrary part of the environment relative to human individuals that a set experience *'will learn something'*: *'it is something of an experience'*, or *'it comes from experience or is dependent on the experience ...'*

The terms "manifestation, statement, testimony" etc. are taken from the ordinary usage of individuals; we keep them because we could be in danger of disturbing theoretical notations associated with these expressions that don't appear to be inclusive enough (*nicht groß genug erscheint*). We ourselves don't associate any other theory with these words. In other cases, where keeping common expressions is less objectionable, we will then not suggest new terms.

2. — To this assumption, the environmental constituents function as a prerequisite to what has been said; and it will be established that the environmental constituent prerequisites of statements then be used as experiences. Next, we will assume that all of the components of the prerequisites of statements are together set as experience. We take the simplest assumption as follows:

Theorem A: As the components of our environment form a composite (all components), so the statements form into experience.

3. — The complete reversal of Theorem A gives the (following)

Theorem B: If a statement with all its components is taken as experience, then also the components of the environment are taken as prerequisites of a statement whose components are then accepted.

Since theorem A cannot be treated as fully reversible without further ado, theorem B cannot be easily accepted. (Author's Note: Avenarius is simply saying that the environment may or may not affect the components that form statements which form experiences.)

4. On the other hand, theorem B makes it easy to deduce an initial conception of **pure experience**: namely the experience is expressed in all the components of the environment which are considered prerequisites for the components themselves. This concept of pure experience may be called **synthetic**.

5. However, in order to arrive at a concept of pure experience, we do not necessarily need to go beyond the scope of the stated concept itself. Suppose there is an experience purely expressed by components, which can be considered simultaneous or original experience which leads to a second form of pure experience: as a statement in which nothing is mixed in, which is not itself another experience — it is in itself nothing but pure experience.

We call this second form of experience the **analytical** concept of pure experience.

6. — Since the theorem B cannot be inferred from the theorem A by a mere inversion (compare no. 8), so the synthetic and analytical concept of pure experience does not simply coincide. If theorem B was true, they would fall in together; and theorem A would be a reciprocator.

I note that where certain expressions remind (the reader) of Kant, it remains completely undecided whether *the concepts* are Kantian or not.

II.

7. — The concepts gained from pure experience determine the general critique of the tasks.

Theorem A denotes components of our environment as a prerequisite of experience. The synthetic concept of pure experience has acquired this assumption; it will be investigated:

In what sense and extent can components of our environment be accepted as **prerequisites** of experience?

8. — In the analytical concept of pure experience, on the other hand, we are dealing with the experience itself, insofar as it contains nothing but experience: a fact or a phenomenon or something else of a certain kind, such as pain or belief is also a fact or a phenomenon or something of a certain kind. In this way we have to investigate:

In what sense and extent can predicated values be accepted as **experience** (*Erfahrung*) at all?

9. — The disintegration of the synthetic and analytical concept of pure experience makes it possible that a statement including all its components is set as experience, without at the same time, accepting elements of our environment as a presupposition of all components. The question arises:

In what way and extent do the synthetic and analytical concepts of pure experience depart and can their **division** (*Zuzammenfallen*) be accepted?

III.

10. — Provided that the attempt to solve those tasks achieves any positive, inter-organically connected results, the entirety of the investigations to be undertaken will take on the character of a theory of experience.

This narrow characterization would, however, be preserved only if the object did not extend beyond itself. But if we already have experience in such a connection – with one of the most important principle values, the knowledge that experience would be a kind of generic cognition requiring us to expand the circle of our investigations to knowledge in general; and all the more because we are not permitted to presuppose any generally accepted, fixed concept of cognition, so that for example, the determination of the differences of types would not be (*Artunterschiedes*) sufficient to determine directly the experience itself.

11. — However, if knowledge were determined according to what was spoken, then the narrower the character of one theory of experience the wider the approximate theory of knowledge. (*Theorie der Erfahrung, Theorie der Erkenntnis* p.7 section 11). Indeed, in contrast to all theories which more or less rest on *justification* based on the school of thought the author has joined — differences among authors and schools of thought — understood as "knowledge" in contrast to all specialized epistemologies, which would be characteristic of one all-encompassing theory of knowledge. This would mean a theory that recognized its general concept of objects (*Gegenstand*).

12. — If however, the connection of experience with knowledge should force us to general-epistemological-investigations, then they should lead to a further extension of our circle of study. It is relatively easy to determine in the sense of a special epistemology: this is the Being (*Seiende*) this is the means of knowledge (capacity, organs), this is the knowledge and so on. It is already – more or less knowingly or unwillingly – that the Being as such, the concept of knowledge etc., is determined by some standard. On the other hand, a general theory of knowledge would have to include general norms according to which individuals exist (*Sein*) and are knowing, experienced and inexperienced, knowing and unknowing, certain and questionable etc. Indeed, even the norm is determined according to its behavior recognizing or acting in relation to surrounding components which at first are only objects of inquiry. — And as far as we touch on such problems, our investigation would become a character of a general theory of human norms.

13. — And now it falls into the area of our assumptions that in particularly important cases individuals do not experience but are in close connection with knowledge re: experience as a kind of knowledge and both experience as knowledge, in a substantial relation to beings and non-beings, true and untrue, certain and uncertain, known and unknown, recognized and unrecognized and so forth. – that is in a substantial certainty according to acknowledged general standards. And so, we must, if we want to just examine experience as such, involve these relationships in our circle of investigation.

But that means, our **Theory of Experience** becomes first and foremost employed in spirit as a theory of experience; that does not preclude rather it requires at the same time a **general theory of knowledge** not fully separated from a **general theory of human norms** (*menschliche normen*).

Regarding the relationship between experience and knowledge together there is nothing at all to be made out – so too not whether all knowledge comes from original experiences just as not all experiences come from original knowledge; or whether knowledge is a sort of experience or experience could be accepted as knowledge. I wish to note only specific cases in which experience is factual knowledge which differs from other kinds of knowledge. Many individuals express a knowledge of God – but as such it is not experience; and appreciate a base form of knowledge. Other individuals who have knowledge of God disagree and say that nothing empirical is added; they seem to attribute a higher rank to certain senses than to empirical ones.

IV.

14. — It will be obvious to try to solve the (nos. 7-9) formulated tasks by an immediate critical assessment of the partial or total justification or non-justification of the assumption from which we proceeded.

Alone, if we do not want to succumb to naïve criticism, we could posit: with which *rights* that assumption was made or appropriated — in which parts it would be *defensible* or *indefensible*, at any rate as long as nothing is decided or as yet determined, what is to be understood about the underlying assumptions in different moments. And in order to determine that, unless we blindly surrendered one of many epistemologies, we again through no other means, decided to simply analyze the various assumed moments. The method, whose application then attempts an answer to our presented question, is therefore the analysis.

V.

15. — One cannot perform an analysis of any kind without taking a standpoint from which to do it. If authors and readers are to arrive at common analytical results, they must start from a common point of view. It then makes sense for me to propose a common position (*Standpunkt*). If the

reader refuses my suggestion because he cannot decide or does not want to share the stated point of view - well, then I must give up the hope of communicating with him for the time being.

16. — As a point of view, I suggest one where the Greek tradition assigns to the business of the philosopher as one who stands amidst the bustle of the market, not as a buyer or seller but as spectator of all actions; he travels through distant lands and associates with foreign peoples not for any higher or lower purpose but to observe.

17. — I mean this point of view literally and locally: on the one hand we have the constituents of our environment, and on the other hand in relation to the human individuals in the same locality like the traveler in an alien place and their people, we are like spectators in the market, or in the theater, like the venue and the audience.

In these investigations, which are the subject of human knowledge, there seems to exist a mysterious compulsion to increase the point of view as far as possible or as deep as possible, as abstract as possible or as concretely as possible in principle. Wherever one arrives at from these certainly exalted and noble points of view show the ragged state of contemporary philosophy. I am far from wanting to change this state. The following investigations will show why I don't cherish such a thing. I just want to apologize if I make a more modest stand for the moment and investigate more modest claims.

VI.

18. — Insofar as the object, function, character, method and point of view of the following experiment can be deduced from the assumption which we have cited in section. 1, this assumption leaves itself as the presupposition of our criticism or, briefly and technically, as the empiriocritical prerequisite condition.

And it was firstly only in the sense that this assumption seemed appropriate to start the undertaking simply and conveniently as a test of experience.

But then also in the sense that at the same time it should embrace all the prerequisites of the implementation of the planned undertaking.

19. — In order to see what the assumption of what is necessary entails and what it does not need to contain, we emphasize above all that it depends on the local point of view, as specified in (num. 17); and that they are components of a particular environment determined by a certain location as well as by human individuals insofar as they can only be thought capable of asserting what they are experiencing.

20. — We stop at the given point of view and stay within our assumption, on the one hand, the component constituents like our surroundings such as plants, or stones, as mountains or streams, as moon or sun, as earth or sky, as animals or even human beings; and on the other hand humans thought to be "ingenious" or "ordinary" as children or savages, as naturalists or theologians, as all-embracing Alchemists or pulverizing Critics. - Not unlike the cogs in a business process or parliament, we stand amidst philosophers, their factions and their strife.

We always accept the following: on the one hand individual humans with manifold statements, and on the other hand the components of the environment as prerequisites of statements.

21. — And then if we take the moments in the sense that they provide a *presupposition* of a critique of pure experience, we take everything that is not marked as mere names or vague phenomenon (*Phantome*) and empty schemes; but the environmental constituents in all their physical and chemical components, we consider too the human individuals in all their components: anatomical and physiological, normal and abnormal, certainty and changeability; and therefore treat as one any certainty and change of such kind as belonging to one all-inclusive empiriocritical premise.

22. — And just as in all differences, like them they are set in certainty and change. So, we take - as prerequisites of our investigation - the different environmental component differences on the one hand in nature, and these again in qualitative and quantitative terms respectively; on the other hand, the arrangement, and this again in spatial and in terms of time; then a third dimension - differences of simplicity and complexity, and then a fourth dimension - differences in frequency or rarity, likeness or uniqueness, in which these recur in the environment.

23. — We take these differences in the environmental components accordingly consider the changes in the latter as being of a quantitative or qualitative nature, the spatial or temporal arrangement and the overall composition; and consider all these changes, on the other hand, as common or rarer, similar or dissimilar recurring, and so forth.

24. — Furthermore, as regards the human individuals we presuppose as proponents (*Aussagende*), we accept them as highly developed organisms, descended from parents and forefathers, fathered by a father, conceived and born by a mother, growing up and living - and dying after a period of conservation. We accept all their changes like psychological processes, nutrition, growth, movement, secretion, reproduction and so on.

25. — And in relation to those human individual statements (*Aussagen*), in which the empiriocritical condition resides, we take the individuals as not mere schemas, their statements not as mere noise and sound, but as words, as expressed symbols for perceptions, memories, thoughts etc., or at minimum as interjections with which the unexpected is related. But we also take the "secretion of the lacrimal glands" known in many cases as crying, that means, as concomitant of a painful mood; the "punctually interrupted expiration" as laughter, that means, as concomitant of a cheerful state; certain movements of the forehead and eyes as concomitant of an observation or distraction, a question or a decision, of understanding or confusion and so on: so that these movements contain the meaning of statements.

VII.

26. — As desirable as it is, the components of our environment and the value statements predicated to individuals, are already distinguished in common terms, which makes it difficult as such expressions already contain habits which can easily be reverted to.

We are confronted by theories that come with conventional expressions of physicists and physiologists, psychologists and philosophers — so woven that the connotations often go unnoticed — that in order to protect us, as much as possible, from the influence of theories, we label any value accessible to description as a presupposition of our environment simply as **R**.

27. — And against this, every value amenable to description, provided it is the content of a statement of another human individual, is accepted and denoted simply with **E**.

28. — How we see and accept perceptions and memories, thoughts and feelings, etc., with respect to other human individuals is as they are contained in the presuppositions in ourselves — generally accepted in other individuals only in the same sense in which we view the constituent presuppositions as part of ourselves.

29. — More precisely, we do not take these values labelled **E** — the E values, as we say briefly — not as rigidly determined presuppositions of our criticism, but changeable and manifold determinants; not only presupposed (*Voraussetzungen*) in the relatively simplest terms but also the most complicated, in the most primitive natural state and highly developed historical forms.

30. — We have so little to provide all the details of this kind, like those on the environmental constituents; but at least one difference from the general assumptions of the E-values should not be overlooked here which is likely to prove valuable and concerns itself with the simplest perception. It is in our empiriocritical premise also to include that human individuals e-values not just expressions like 'green', 'blue', 'cold', 'warm', 'hard', 'soft', 'stiff', 'sour' etc., as well as terms like 'pleasant', 'unpleasant', 'beautiful', 'informal', 'pleasant', 'disgusting' etc. If an E-value is specified in more detail and can be expression by expressions of the first kind ('green', 'sweet', 'Sound a') from here on it is referred to as an element. If it can be specified by the name of the latter kind, we call it a character; and for the time being will reckon with the characters as analogous to E-values in the same way 'pleasant', 'unpleasant' etc. as the other E-values are set like 'color', 'Tone', 'smell', 'taste' etc.

VIII.

31. — We have only to take a look at the surroundings, provided that it presupposes any determinations or changes in the human condition. Our general empiriocritical presupposition comprises such a relation of the environment to the human individuals in two ways: the first already expressed in (sec. 1), the second results from the presuppositions which we presuppose when accepting human individuals.

Regarding the way of the former, it contains not only that when any environmental constituent R is set, for instance, any experience or otherwise any E-value is stated; but also, in the case that if - for example in any chemical or physical experiment - variations of the R-values are set, the statement of the human individuals in turn follows with variations in the E-values. In such cases our empiriocritical presupposition encompasses two series:

1. $R, R', R'', \ldots R(n)$
2. $E, E', E'', \ldots E(n)$

their members in one environment comprised of themselves, in the environment of other individuals and it includes at the same time that as they follow the variations in a certain way, the members of the second row E, $E', E'',\ldots,E(n)$ are somehow dependent on the members of the first row R, R', $R'', \ldots, R(n)$ (*abhängig anzunehmen sind*)

32. — Just as the values of the second series depend on the first in our empiriocritical requirement and as earlier, on the one hand, the environmental components R determine the temporal and spatial conditions etc., on the other hand, the human individuals are assumed to contain physiological and other types of determinants (*Bestimmtheiten*); so our general requirement also includes specifically the assumption, that the E-values - in their dependency on R-values - depend on the nature of R, the duration of time in which the conditions are set, the distance in which they exist from the individual, depending on spatial and temporal arrangement,

like physiological circumstances in which the individual, as a result of temperature, air pressure and the like, is able to be modified.

33. — Furthermore, our general empiriocritical assumption includes the following special assumptions:

a) if the same R is set at different times, then one does not have to assume that for each time one and the same E-value is in dependency (*Abhängigkeit*)

So, a child may consider a circle one time as a plate, and in another instance as a moon; a square once as a candy and another time as a table. The R-value remains for a longer time, so sometimes different E-values follow immediately. A figure that was extremely problematic, which was supposed to represent a tree stump, later became a tree stump, in the artists hands but was considered by another as a Mercury wing helmet from another unprejudiced viewer it was initially seen as a pig and in the second instance as a hat. If a word is mentioned it may have one understanding then a second, a third ... Similarly, with personal names different individuals having the same name may differ spiritually. In one case an individual heard - a noise in the first moment as distant cheers of a sleigh-rider and in a second moment the snow falling from the house; another individual first heard the one in the same noise as a domestic dog hunting a cat. In another case an individual saw something white in a window facing his apartment; It developed the following series of values, expressing what it could be: a curtain; white paint; pre-glued paper; reflected light.

34. — If R-value is set, an E-value in an association is accepted, not assuming that the value set is only one E-value; rather with the same R, different E-values can be set and it can be any E-value as an initial member of a more or less large series $E(1)$, $E(2)$,...$E(n)$.

At an E-value conditioned by a certain R-value, there are other E-values more external, meaning they exist without being an immediate determinant of what is seen, heard, etc., to continue: you think, for example that the E-value first set is similar in an image you see, in a name you hear, in an anecdote that you read and so on, especially in pictures, names, stories, etc.

35. — If R is presupposed and a series $E(1)$, $E(2)$, ... $E(n)$ must be assumed, it must not be assumed that the individual members had to be related to the first R element; but it can also be that disparate elements can occur so that no E-value can enter an existing R set. The sight of a particular fruit may cause a distinct sensation of taste (e.g. sour) and inversely, a sensation of taste exists in the idea of fruit. a piece of music that is heard may contain the notion of a room in which it was

once played, and in the sight of that room, inversely is contained the idea of a piece of music. Odors can be present in certain life situations in which they were once perceived and vice-versa, certain life situations can be related to particular odors. A drunkard may feel the slap of a passing lady for some time afterwards. In a word: every element of the sensory area can exist with every other sensory area and therefore are associated according to ordinary psychoanalytic logic.

36. — It follows from this,

1.   that the E-values which, when an R value is set as dependent of it (num. 83) or as a connection to these dependencies (no. 34 and no. 85) can represent a plurality whose elements arrange themselves in a row;

The number of E-values, which when the R-value is set, is conceivably an indefinite quantity, which in reality can be accepted as a choice from the indefinite number of conceivable ones. In this case, the E-values available for selection are more diverse and conditioned by the more varied R-values and the development of the individual (through more advanced teaching, richer experiences etc.,).

I take the expressions "thinkability" and "reality" here in the sense of ordinary usage; a closer determination of this same follows.

IX.

37. — But we do not just accept the prerequisite of realized experience and other related E-values. On the contrary, just as we assumed human individuals as being understood in the process of development, etc., (num. 24), in another sense we have to assume the environment in especially, in particular as a presupposition of individual preservation: as far as the environment must be thought of as food and shelter.

Nevertheless, we as human individuals apart from the entire environment must presuppose existing in an environment that guarantees absolute preservation. And if in turn, we consider only one environment contained in our condition, which is not without hazardous and harmful elements, the assumption also includes the ability of the human organism, itself experiencing damages and diminutions, to preserve itself within certain limits - and in such cases, where the environment alone is not at the same

14

time causing immediate damage or defending against it.

X.

38. — The preceding remarks have sufficiently unfolded our empriocritical presuppositions to show, in the most general outlines: everything necessary for our critique of pure experience is inherent in it. It will be a matter of the following analysis, wherever its progress requires it and as much as our purpose requires to emphasize the additional and special.

But only then and nothing more. For now, all that remains is to briefly indicate what our assumption does not contain and what is not necessary:

The empiriocritical presupposition is - according to the idea (*Idee*) - to include all material, from which the philosophical systems and special epistemologies develop; but - according to the ideal (*Ideal*) - nothing for which system and theory first make of it.

39. — That means:

The empiriocritical premise does not add any further prerequisites to the terms which the analysis of its individual assumptions would suggest. He who makes the empiriocritical presupposition is bound to nothing, assuming further as with the notion of R values as well as a concept of "matter" or of "object of knowledge" or "thingness" or even whether "substantiality" is assumed or not; or with the concept of E-values already some "soul concept" or just a concept of "consciousness" or "reality" κατ' ἐξοχήν (most importantly).

The concept does not connect to an immediate assumption about the "possibilities" of ratios of E-values to R-values in kind and scope; or with any concept of dependency like "causality" or "necessity" or "freedom"; or in general with the concept of our presupposed concept of "being" or "appearance", "reality" or "ideality" and so on.

15

However, as the following inquiry does not require any further assumptions, it must also reject any liability for consequences which might be brought to it by far-reaching conditions.

PART ONE

OUR ENVIRONMENT AND THE SYSTEM C

First Section
Our Environment

First Chapter
Common Terms

I.

40. — The first task, which derived from the concepts of the latter for a critique of pure experience, led to the question: in what sense and to what extent can components of our environment be accepted as **prerequisites** of experience? Since the first question asked is based on the presupposition of experience, and provided that this assumption remains part of **our** environment, to answer we have to subject components to an analysis of the totality of what we consider, from our point of view (*Standpunkt*), as our environment. Before we can do this, we must agree on some more general terms.

41. — Let us think of two of those in our general empiriocist assumptions that variables V_1 and V_2 are indifferent but in any case, so linked that when V_1 changes, changes are set for V_2, thus we associate V_1 as a condition of changes (*Änderungsbedingung*) in V_2; the changes on the other hand of the second variable V_2 are narrowly conditional or dependent on V_1; and finally, we deal with both variables under the concept of a system.

42. — Insofar as all the constituent parts of our condition are thought to be changeable and their changes dependent on each other in the manner indication previously, we believe them to form the most manifold system variated in size and joined together in a single all-encompassing system, which we provisionally call System R.

43. — Regarding any system of environmental constituents (*Umgebungsbestandteilen*), provided that it only satisfies the condition that it forms with another variable a system of higher order – that is, even a variable whose changes depend on another variable -, we denote the totalitality of the features through which the individual concept of the system would be logically completed at a point in time as x_1, and as the systemic condition of a time point x_1 (*Systembeschaffenheit des Zeitpunktes*).

44. — I think of a system C changed in the following time as x_2 and refer to the system as it was originally set immediately before the change, as the initial condition of the system (*Anfangsbeschaffenheit des Systems*)- and as it was set immediately after any change, this is referred to as the end condition of the system.

45. — A change e.g., in any variable, including the system variables, cannot be thought of more simply than that the variable is thought of on the one hand at least two times – in an earlier state τ_1, and later τ_2 – and on the other hand the value τ_2 is thought to be added positively or negatively. For our purpose, therefore, the simplest way of understanding the term ,system C change (*Änderung eines Systems*) is to understand the positive or negative addition of that particular quantity. We denote the system only by V, then we obtain the analytic expression $V + \Delta V$ to denote the final condition of the system.

46. — A change, ref., an end condition of any system which is set in the same sense as part of the empiriocritical premise, such as the movements of my pen writing, I recognize as real.

47. — If, after a change of the system at the point in time τ_2, I set the system back to its original state before the change — that is point τ_1, which remains in a logical relation to the change, I designate the system C changed in τ_2 as changeable in relation to the imaginary change. And the relevant - as we will say - reinterpreted change, e.g., I call it the final condition that belongs to it in relation to the system.

48. — If I multiply the presupposition of a system by the presupposition of a condition of change at all, then I designate every system C change is presumed in advance of this kind, viz. Final condition, if only its concept does not contradict the general concept of the conceived presupposed system itself.

49. — If at the point of change in τ_2 e.g., the final condition of a system, which I have designated for time point τ_1 as possible or conceivable, so I name this change as well as the end condition set with it, as being realized, e.g., as the realization of the conceivable or possible change, aka the end condition.

50. — It requires, however, consistency in conditions, a change that I call the necessary end condition that has been realized.

The terms introduced should be given no other meaning than a logical meaning.

51. — By changing time, I understand the time it takes the system to make the conceivable change. These change times, as noted here, are not thought of as immutable for all systems. An assumption of this kind would contradict our general empiriocritical presupposition.

II.

52. — Assume that the symbol $V + \Delta V$ does not designate a change in general, but a clearly determined concrete change of system V which was also set at the time point χ_2 when the uniquely determined environmental constituent R was set. If we now return to the time point χ_1, that change of V on which the final condition $V + \Delta V$ was based on remains possible; V itself assumes the significance of the totality of those conditions under which the change in question, insofar as it depends on V alone, can justly be described as possible.

53. — But from this it follows:

If I designate a particular change of V as *possible*, then I cannot call the same change as *real* as long as I limit the sum of its conditions to those contained in V alone.

If a particular change, after being designated as possible in t₁ (according to the preceding one), should be called true in t₂, then the sum of the conditions contained in V alone must be increased by at least one; which was previously not included in V.

54. — The co-condition which is required inside V, must be added to those contained within V to indicate the *reality* of *possible* change, and we call this the complimentary condition of the corresponding changes; the conditions contained in V in their prerequisites. — In the assumed case, Rx denotes the complimentary condition.

55. — We call the composition of the systematic preconditions and the complementary condition a conditional whole. (*Bedingungsgesamtheit*)

56. — Hereinafter, a change *possibility* maybe called *true* only in the case that not one or the other of its presuppositions is set, rather the conditional whole is set.

III.

57. — From nos. 44, 45 we have results for the final condition of a system:

If the change of an initial system is thought to be a consequence of the setting of a change condition (*Setzung einer Änderungsbedingung*), the end condition cannot be determined by the change condition alone but must also be determined by the initial claim intended to be thought of regarding the changed system.

Second Chapter
The Components.

I.

58. — If we begin to dissect everything contained in the empiriocritical premise, we next consider our point of view and purpose and place on the one hand the relation established in no. 20, which is considered as the following statement: fellow human beings or humans as such – are referred to as the individual.

We consider him with everything that can be thought of in such a relationship with individual that, when it is set, and changes in that individual are also set; inclusive of everything that can possibly be considered as a condition of change in relation to the particular individual.

59. — We don't think of anything else that according to our requirement and cannot be thought of as a condition of change with regard to the particular individual, and which can be considered meaningless for him and us.

The totality of all that is meant as a condition of change (*Änderungsbedingung*) in relation to the definite human individual, we designate, in agreement with n. 20, as the environment of a given individual, or the individual environment.

In the course of our investigation, when we speak of environment in general, we mean the *individual environment* - that is, our environment, as long as it coincides with the environment of the individual.

II.

60. — We can now subordinate the environment of the individual into one of our important divisions – two parts, accordingly as making the classification from a - say "pedagogical" or, say, "physiological" point of view.

61. — In the former respect, one then obtains the difference in the individual environment as "locality" and "social circle".

62. — In the physiological sense, the environment is conceived in a different sense as a condition of change for the individual, when its constituent parts enter into the relation of a *subject matter* or a *nutrient*. Accordingly, we also re-allocate the environmental components - not in themselves, but in their relationship as a change condition for the organism; and under the designation S, here we are concerned with everything supplying the organism from outside, its metabolic forms and at the same time we use the assumed symbol R to designate everything that, in its general notion, in the language of physiology then, as „general or specific stimulus exciting a nerve."

There is nothing in the *environment of the individual* which cannot change the organism in the manner indicated, although it does not necessarily need to change it in the manner indicated.

For example, mechanical pressure or shock is a "general stimulus", but it may also crush the organism in some circumstances. "Chemical action" is also a "general stimulus", but it may poison the organism under certain circumstances. Even "specific stimuli", such as light and sound, may destroy the sensory apparatus in question. As nothing is presupposed in the individual environment, which cannot "excite" a nerve, I would also like to attribute to the individual environment all that thus changes the organism, even if it has received its momentary place within the organism, i.e., the latter case as a so-called "inner", and thus considered as the "main stimulus".

III.

63. — According to our assumptions on the one hand and our concept of the system (no. 41) on the other hand, every particular human being is composed of a plurality of subsystems (*Teilsystemen*) and in this composition of subsystems as a whole is to be regarded as a system itself. So, we have to further dissect the notion that a particular human is a system of subsystems. In order to gain a point of view for our purpose, we pursue one side of the empiriocritical presupposition. In fact, the assumption (no. 31) contained within it can be formulated as follows

Theorem I: In some cases, when R is set and E is assumed, E is also somehow dependent on the acceptance of R.

64. — If we consider the theorem referenced above, I think of the value E as dependent on an environmental component R; I cannot think of E as directly dependent on R. Because if, for example, an (alert) individual picks up a vibrating tuning fork, they would not be able to testify to an associated E, termed 'sound', if the auditory nerves were destroyed at their periphery or central end or in their circumstance. Likewise, if the retina of his eyes were annihilated, or the optic nerves cut, or the central receptors degenerated, the individual would sense no 'color'. Likewise, if the skin or the nerves or their central endings were destroyed, no 'E' or 'coolness', for example, would be assumed.

65. — Likewise, no movement of the individual would take place if the peripheral connection of the motor nerve with the muscle or the motor nerve itself were interrupted, or if the organism itself, such as through bruising, would have suffered a major disturbance.

66. — Since it is precisely these subsystems which may almost exclusively be considered for our purpose, we must be free to undertake such an analysis of the total system C called „man", as it is most useful for our purpose.

Accordingly, for the time being we distinguish only two things about the human individual:

I) The nervous system.

II) The entireness (*Gesamtheit*) of the remaining subsystems.
The further decomposition of II, if needed, will be borrowed from the special sciences; For the further analysis of I, we again provide only for our purposes and only as far as it requires.

IV.

67. — The nervous system, although in itself a subsystem, we imagine in the sense of anatomy and physiology, again as composed of manifold subsystems of higher and lower order. It is sufficient for our purpose to simply distinguish between nerve fibers and the central structures: the usual further division of the nerve fibers into centrifugal, centripetal and intercentric, centrifugal into motor and secretory, centripetal into sensory and reflexive – here mentioned to show a difference to be made within the most important central system, the brain.

68. — For as E is not directly dependent on the surrounding matter R, it does not, in general terms, depend directly on the extreme peripheral end of the nerve; for cases of E can also be assumed without the participation of peripheral endings, for example, the so-called "sensation" of amputated extremities. However, E values are not necessarily directly dependent on the nerve fiber; because we can again assume such values exist when the relevant nerve fibers are eliminated, such as in the case of so-called facial hallucinations in the atrophic optic nerve).

69. — So if I follow such a nervous structure from its extreme peripheral end, through the fiber and into the brain, and further, I reach a nervous subsystem, from which E directly depends, meaning that I could no longer accept, as an (experimental or pathological) omission, without also having to accept the dependent E – it's very dependency on E – as well.

70. — As the stated central subsystem is thought of as the place in which all the endings of centripetal nerves, insofar as E-values directly depend at least on these endings, are united, it must also be thought of as the place where all the centrifugal nerves originate - as far as at least their function, which continues in muscle contraction or gland secretion, and at its central end — in temporal relation to E-values.

71. — This nervous subsystem, which collects the changes emanating from the periphery and distributes the changes to the periphery, seems to me to serve as a distinct concept for the purpose of the comprehensive system of central organs; While I leave its closer anatomical and physiological determinations - as not so sure of their assumptions at all – isolated from our aims, it can be left undecided. I simply call the adopted subsystem the System C.

It therefore results in the decomposition of the nervous system for our purpose:

A) the System C

B) the rest of the nervous system

V.

72. — For a further distinction with respect to the system C, the point of view could be derived from the assumption that a single associated "function" does not claim the entire System C for its operation. The functions derived from this presupposition determine the definite parts of the system e.g., to find particular functions for given parts of the System C leads to the attempt to transfer each function to a delimited area and thus divide the organ itself into spatially juxtaposed „centers" or „spheres" and the like.

Reflecting on these divisions is not necessarily our concern; they are more important for brain anatomy and physiology than they are for us, and moreover, to this question, the individual theories are both general in their view and, in the latter case, hostile in their "observations."

If, however, cerebral physiology seems to be burdened with contradictions - precisely in the spatial determination of the parts of the central organ to be attributed to the various functions, in any case, this spatial division, limitation and design is out of the question for us, but it is certain enough for our purpose, if only for the general presupposition that the division of System C is based on certain functions which are held together without further determination of a spatial nature.

73. — We are satisfied with the general requirement: — no matter how great — the multiplicity of form elements — cells, neurons — they have adopted a certain change in the process of division of labor; these may now have the meaning of a motor, secretory or sensitivity function. We express this as: a multiplicity of central forms - elements functionally connected in a certain sense; and we name every such connection with definite meaning a central partial system.

74. — For the time being, the meaning of the central partial system should therefore be primarily a functional one; not a spatial one — although the central partial system is ultimately just as spatial as the System C itself. And if we sum up the central organ C, as far as it is even an *organ*, in accordance with what has been said, as a totality of central partial systems (*zentraler Partialsysteme*), we leave it open whether the latter are more precisely stored, juxtaposed or mutually interspersed, with or without common form elements demarcated or scattered, etc.

75. — If one thinks of two or more partial systems as functionally connected, then such partial systems of higher order may be called, for example, coordination systems, which can then reunite into coordinated systems of a higher order.

76. — Accordingly, within these partial systems we would then characterize those which are distinguished by greater formal and functional development from others of the same system C. We provide for this difference with the terms main and minor partial systems.

77. — If the first distinction is based on the composition of the partial systems, and the second on their formal and functional development, then, finally, the same difference must be made in regard to their functional relationship as we have in the fibers (no. 67) (*Nervenfasern*) and also in the partial Systems themselves (no. 73) having mentioned this in passing:

Depending on whether we think of a central partial system as an E-value, in the sense of our assumption, we are directly dependent on it or on its change into a movement, in other words: expiring in a secretion, obtaining a sensory or a motor movement, e.g., secretory central partial system.

78. — Another difference is made for us within the sensory partial systems, with changes based on an environmental component, as it approaches the peripheral end of the apparatus from outside the organism in a specific manner (such as sound waves to the auditory nerves or gaseous substances to the apparatus of the olfactory nerves).

VI.

79. — The above can be found in two sentences. After the condition of the system C is obtained, the content of the (no. 68) can be expressed as follows:

Theorem II: In any case in which E (*E-Werte*) is assumed to be dependent on R, E is assumed to be directly dependent on C.

80. — But since it follows from the same presupposition, from which theorem I (no. 63) followed, that, unless the members of the series R_1, R_2, ... R (n) were set, the members are not the series E 1, E 2,. , , E (n), it follows that E depends only on R if the substitution of R caused a change of C.

Theorem III: In any case, in which E (*E-Werte*) is assumed to be dependent on R, E is assumed to be immediately dependent on a change of C.

Chapter Three
The Changes.

I.

81. — According to theorem III, and according to the empiriocritical view, we treat the relation between E and changes in system C as independent, and E as dependent.

Just as we made further assumptions (see no.39) regarding the dependence of E and R, so here we recognize the dependence of the system C change based on the R-value as well as the E-value.

82. — Since, according to our presupposition, the environment and the individuals themselves are assumed as variables, further decomposition will be directed to the examining the types and quantities of their changes insofar as they are suitable. According to the division of our whole into an individual environment and the individual (nos. 58 and 59) we now include changes in:

I. Changes in the environment and
II. Changes in the individual

83. — The environmental changes (I) are classified in the same way as the specialized sciences, the movements and textural changes in the environmental components (*Umgebungsbestandteilen*).

84. — The great number of changes in genus II, that is, of the human individual as a whole and in its subsystems, enumerated by a comprehensive classification, is not our main concern. We limit ourselves, in the sense of our task, to the following classification:

A) changes of such non-nervous subsystems and system parts, which are somehow represented by a nervous connection with the system C and thus also exist in a dependent relationship with the same;

B) changes in the nervous system itself

C) changes that don't belong to A) or B). —

In refraining from the changes in type C), which are not required for our purpose, we shall seek to establish from the changes of type A) and B) —as far as is important for us — a division which best suits our purposes. For completeness as well, the absolute limits of the distinct elements should be claimed.

85. — The changes of the type A can be divided into subtypes:

1. Changes with which a change of place is connected; for example

a) A change of place of the individual and
b) Change of place of the environmental component — or
c) Cancellation of subtypes a and b

2. Changes connected with a change in the environmental component; in fact

a) Dietary intake
b) Mating (including fertilization)
c) Repeal of existing changes, in the sense that
    a) conservation (care, protection)
    b) destruction (damage, destruction, injury, killing)

d) Alteration

3. Changes without simultaneous changes of location or surrounding components; in fact

a) all so-called physiological functions of the organs represented in system C, which do not serve those mentioned under 1 and 2; for example, accommodation of the lens and tympanic membrane, constriction and dilatation of the pupil, secretions, changes in cardiac and respiratory activity, etc.;

b) Heat (and Electrical) development;

c) Digestive and nutrition processes

We accept such changes individually, in groups and in series. If we assume repetitions of more or less composed series of such changes dependent on C, then their composition cannot be assumed to be unchanged in any case.

86. — The changes of the type B, of the nervous system, we separate first for our purposes:

1. Changes in system C.

2. Changes in the remaining nervous systems.

Dividing further these last two noted changes is far from our task.

II.

87. — Turning now to the disassembly of the changes in C itself, we seek to distill a guiding view from our assumptions. We remember that sentences I-III (nos. 63, 79, 80) contain the values R and E existing only as very abstract terms: what was mentioned of them was that the symbols R and E did not represent concrete, but rather simple genuses (*Gattungen*).

88. — If we take R as the *starting point* for a differentiation of the changes of C, we have two paths: We can either vary the concept of R again by all conceivable determinations and thus decompose the concept of a change of C into the individual concepts of all changes of C dependent on the determinations of the concept R; or we reflect the general changes of C, which if any subsumable case is given under the notion R, as a condition for the assumption of E, then it must be given in C that E can be accepted.

89. — Since the latter path may be more in the sense of a *general epistemology*, the former may be committed here only insofar as it leads us to a distinction which we already made at the very beginning (no. 62): the distinction of the environmental components as change conditions for a particular human individual — and thus for C — in the two classes R and S. Denoting the changes of C, if they depend either on R or on S, with f(R) and

f(S), we have distinguished the major classes of changes of C by their dependence on R.

III.

90. — If we now take the other path to obtain the modes of change which, when R is set, must generally be set in C, so that E can also be accepted, then we proceed most certainly if we distill the guiding point of view for those changes claimed in C with the proof that their claim is a legitimate one.

For this purpose, it is advisable to emphasize the abstractness of the sentences I – III, while at the same time use our proposed terminology as needed.

We therefore summarize the result of our analysis as follows:

The variable system is assumed in C in which the E-value is assumed to be directly dependent on its end condition $C + \Delta C$ determined by R.

91. — It follows:

The given case Σ depends directly on $C + \Delta C$, and indirectly on R as the condition of change of each end condition. And it is this end condition $C + \Delta C$ thus accepted as the immediate, and R as the indirect condition of E. Hereby the dependence of even the E-value from R and $C + \Delta C$ is only assumed to be logical – meaning – the assumption of this independence contains nothing, except that if R and $C + \Delta C$ are assumed, then Σ must be assumed.

92. — If one takes the proposition that Σ is conditioned by R and $C + \Delta C$ (as E is taken, because with $C + \Delta C$ it behaves quit abstractly), then the two conditions as well as their conditionals must be quite abstract - meaning – in general these are taken as generic.

It may therefore be that so long as the general conditions have not been complicated by the addition of special ones, so the content of the concept Σ

may also be accepted only in a very general way, which in turn does not mean that the content adopted is itself also a general (abstract) one.

93. — On the other hand: If the content of concept Σ is to be thought of as a very special one, then the general conditions must also be duly complicated by the addition of special conditions.

94. — The specified requirement is satisfied in relation to R as R is represented by any part of the environment, in any case, - if one wants: by *showing* a clearly defined environmental component.

95. — To satisfy the same requirement for $C + \Delta C$, as necessary for our purpose, we remove any special case of the general- empiriocritical premise. It is the case that when a specific environmental constituent Rx is *pointed out* (according to what has been said here), a very special „value of Σ ", denoted as Σx, is assumed; we call the corresponding end condition of C once more $C + \Delta C$. Assuming Rx is taken at timepoint t_2, and $C + \Delta C$ at the same time, it is assumed:

the instant immediately preceding the instance of Rx is t_1. Thus, at time point t_1 the end condition $C + \Delta C$ was the only one *possible* (no. 47) C, however, represented the totality of those conditions under which the change in question, insofar as it depends on C alone, can justly be described as possible, that is, the entirety of the systematic preconditions; and finally, R_1 is the condition by which that set of systematic preconditions had to be increased in order for $C + \Delta C$ to be *realized*, that is, the complimentary condition (*Komplementärbedingung*) (no. 54).

96. — Now, however, the general empiriocritical assumption can be inferred, that, if any end condition of C at a given time, that t is complimentary due to Rx and Rx has remained unchanged at an earlier or later point in time, however, the conditional relationship between Rx and t does not have to be thought of as unchanged at the earlier or later time point.

97. — Consequently:

If we kept Rx unchanged but think the condition ratio of C to Rx changed, then we must also think C changed.

98. — Due to the change in C and therefore the conditional relationship of C to Rx it is conceivable, that a particular Rx can cease to have the meaning of a complementary condition in a particular case, or in practice for a special change in C to be the complimentary condition; but even so for example a special R value, which at a certain point in time was not yet a complimentary condition for a certain change of C at a later time or for instance C assumes the meaning of the complimentary condition.

99. — In this way, on the one hand, our notation exists for the changes, e.g., the final state from system C to R values as complimentary conditions. This means we may designate certain R-values under the same conditions as those with *real, possible* or *conceivable* complimentary conditions.

100. — On the other hand, it turns out that C, as the epitome of systematic preconditions, is not presupposed as a variable without history or development, but (it) contains conditions which are reached or abandoned by changes.

101. — These changes, beyond which system C ceases to be the set of systematic preconditions for a particular E-value, if they remain constant; then by adding them, C becomes the epitome of the systematic preconditions for E, even if R remains constant — thus the concepts seem to supply those particular conditions with which the notion of system C, as the *general* condition, must be complicated for a particular E-value. Here then is the guiding principle for the search of those changes in C, which we have to highlight for our purposes.

IV.

102. — Such changes in C, which cause or become a complementary condition Rx for a certain other change of the same system at a later time, are called preparatory changes.

Within them we name those changes on which the difference of waking from sleep is physiologically based, as the general-preparatory changes; and contrast them with the special preparators which come into consideration only with respect to the system C placed under conditions of awakening (*Wachseins*).

Under *sleep* I am permitted, in the sense of my task here, to understand only dreamless sleep — peculiar changes of C, which do not allow the assumption of special E values when R is set, although otherwise for example, painful injuries and inflammations, or deep mental shocks that would create conditions for the adoption of E-values.

Transient forms of the conditions of sleep and waking appear then as the dream, as hallucinations immediately before falling asleep, and so on. — As a limiting case between the general physiological and the special-preparatory changes, we are able to vary the value of R as a complementary condition, fatigue then can be classified.

What is left is the peculiar changes of C during *fatigue*, sleep, and wakefulness.

We ourselves reflect on the remainder only insofar as it is meant to exist under the conditions of being awake.

103. — The special preparatory changes can then be further subdivided into pathological ones, which are assumed to be dependent on temporary or permanent *anomalies* of system C and in the physiological ones, which in turn depend partly on the *activities* of various kinds (*gesetzten Übungen*) set throughout life, partly on *normal developments* set only in certain periods of life.

The question, of whether and to what extent the changes made in the typical course of development can ultimately also be understood as an exercise no longer belongs here.

104. — In order to short-circuit the latter in advance for our purpose, our general assumption as such is *typical development* of growth, puberty, involution, and senile processes of regression.

In the following continuation of our simple decomposition, we will now focus on *activities* rather than physiological changes throughout life.

V.

105. — The emergence of the special-preparatory changes cannot be done at one time (at a time) but must be presumed as distributed over a multiplicity of successive moments of time. In any case, it must be granted the time that C has spent *developing* the central organ as described by the relevant special sciences; because it is precisely the *developmental differentials* that finally establish system C.

But the development of the system C does not begin after or with, but before the birth of the particular human individual; and it will, in the last instance and due consequence, have to be thought so far back, until a development *capable of the development of* C is accepted, which itself can be thought of as nothing further (*nicht weiter*) than something developed earlier.

(*) Vgl , Richard Avenarius Bemerkungen zum Begriff des Gegenstandes der Psychologie. Vierter Artikel. Vierteljahrsschrift f. wiss. Philos. XIX. 1895. p. 136

106. — By distinguishing the preparatory changes made before and after birth, we find a second division between the innate and acquired. The innate would be distinguished between the inherited and the contingent; among the inherited it is understood that there are those whose pertinent change conditions are thought of as ancestral, and those that are not considered so.

107. — And finally it would again be necessary to distinguish between the parental and pre-parental changes within the inherited preparatory changes; in that we understand among the former only those that are inherited, whose corresponding change conditions are thought to be of direct descent

— and among the latter whose conditions of change are considered to be of indirect descent.

108. — By combining both distinctions, we now get the differences of the innate, re- inherited and acquired experience (as well as the innate, inherited and acquired anomaly).

VI.

109. — Within postnatal physiological changes, we distinguish the importance for the function of C — always: for our purposes — between total and partially temporary. By a *completely momentary* change, (*ganz vorübergehenden Änderung*) we must understand one instance when R is cancelled, leaving system C exactly as it was presupposed before the change made with R. We call such change – following our earlier usage of the term — functional or, in short, a quintessential function.

110. — On the other hand, under a *change which is only temporary*, we understand that even if we think of R as cancelled, it nevertheless puts back C as more or less intensively and more or less constantly changed. Such a change is called formal (or organic); remember that in relation to changes it is not necessary to think of the whole of system C, but only that part of it concerning the associated functional changes which can be thought to be limited — i.e., limited to the central partial system.

Whether changes are required, which in the strict sense, leaves a partial system just where it was before the intervention of the change condition – is, an open question remains whether functions are contained in the absolute sense of the breadth of our general presuppositions (*allgemeinen Voraussetzungen*).

111. — We refer to the remnant of a temporary change as remnant change (*Änderungsremanenz*); a change containing a relatively small remanence as relatively volatile, and a relatively large remanence as a relatively sustainable change.

VII.

112. —We gain a further division of the functional changes if we pay attention to the quantitative relationship to the totality of conditions. If we think of a determinant W and a different quantity U as a conditional aggregate, then W must not be thought to be greater or less than U; not smaller, because otherwise U would not be the conditional whole; not larger because then the addition no longer would belong to the conditional whole.

113. — If therefore every condition must be thought of as equal to the total conditional, then in our case the conditional, namely $C + \Delta C$, may very well be thought to be greater than the conditional R; However, if one thinks of a change of the system required to set $C + \Delta C$ greater than R, one must keep in mind that R may not be taken as a condition, but only as one of the conditions involved, namely as a complimentary condition. In this sense, even if, for the purposes of our decomposition, we uniformly refer all changes to R, we can distinguish the functional changes of a partial system into those considered R and those which we think are greater than R. The former we name change equivalents, the latter are considered changer triggering or quintessential triggers. (*Auslösungen schlechthin*).

114. — On the one hand, the triggered changes are on the one hand divisible into peripheral and intracentral; on the other hand they can be described as secondary — in contrast to the triggered systematic changes considered primary — which are therefore to be presupposed as the complimentary condition for triggering.

VIII.

115. — The formal (or organic) changes differ from those considered as dimensional and weight increases and those considered as form (ie., reshaping) the internally constituted parts of system C. The former would be quantitative, the latter constitutional.

116. — If the acquired activity can be thought of as a continuation of the innate, any acquired action can also be referred to as activity augmentation; and as positive, if the associated changes are repeated – as negative, if they are not repeated (activity cessation, lack of activity).

117. — Overall, where further training of system C is based on *activity*, it is thought to be a positive augmentation. Accordingly, regression (degeneration) of certain system parts (form elements), insofar as it is presumed based on activity (labor) deficiency, is thought to be based on negative activity augmentation.

The fact that we hereby do not presuppose activity as the sole condition of positive and negative growth, such as training, is already made clear (no. 104): the typical recording also include the last-mentioned changes in the central partial systems.

118. The division of labor, which we (no. 73) already used, can be derived from the initially stated assumption that the practicing environmental components differ in quality and quantity (no. 22) and the added assumption (no. 109), that activity co-determines, to a certain extent, the formal and functional certainty distinguishes the partial systems from each other. Here, however, we must only stress the importance of a moment that is important for our consideration with regard to the changes in system C at all: this is the prerequisite for different forms of change.

119. — Let us therefore think of the formation (or reorganization) of the central partial systems as progressive formal and functional determinacy of the specific activity, and thus still dependent, to a certain extent, on the specific determinateness of the practicing moments themselves; In addition, we find in our general presupposition moments which, within their qualitative and quantitative differences, are at the same time more or less alike — in contrast to the active moments which are thought to be standing outside any relationship: — so we also assume that the system parts activity by related change conditions in turn acquire related forms of change. We can then divide the forms of change generally into those which are more or less related to one another; and those that are not.

120. —Inside the form similarity we can again draw two limited border cases of the closest kinsfolk on the one side and the other side by the designation for the furthest side: The *relationship in the narrow sense* — and for the other side: a *contrast of the narrower sense* of each then the two extremes

communicated through all sorts of gradations or transitions (*Abstufungen oder Übergänge*).

121. — The difference between the functional and form determination of the partial system: If this certainty is assumed to be dependent on the difference of the active environmental constituents, their settlement is also thought to be dependent on the differences of inherited practices (ie, "*innate disposition*"). And also, the differentation of the forms of change.

122. — In contrast, ancestral activity (referred to as "innate") is thought to be excluded, for the functional and formal determinateness of a partial system and for its form of change, those manifest expressions which, when compared with others, are to be thought of as the determinants of the larger active quantity. (*das grössere Übungsquantum*), therefore those which are set most.

123. — As much of the form and function of a partial system must be considered, so system C must be thought of as the diverse activity of its partial systems (*verschiedenartige Übung seiner Partialsysteme*). The way in which the central partial systems are assumed to differ in form and function through differently weighted activities, they also must be distinguished in their relative importance for the entire formal and functional value of system C. The difference of systematic importance as we shall call it, is based on the most active and therefore the most developed partial systems from the less active and less developed; the latter difference which is already distinguished between the principle and subsidiary systems (see no. 76). From this follows the distinction of the changes according to the systematic meaning of their associated partial systems.

124. — Finally, a sentence may be noted with the results from our last deconstruction and conclusion (no. 57).

The content of the assumption of E-values are made regarding individuals; in a particular case, a specific value Ex is used, thus its setting is thought to be directly dependent on a final state of system C and indirectly dependent on the complimentary condition Rx — the associated form of change is thought to be conditioned not only by the specifically determined condition, but also in the specially determined preparation of system C (e.g., the central partial system).

IX.

125. — In addition we must take from our general presupposition the following special case concerning the complementary condition, in order to complete it by our decomposition of the changes of C, without thereby simultaneously modifying our results. Assuming that a certain environment component Rx implies a change of the central partial system c_1 of C_1, the change continues on to the partial system c_2, whose changes in a movement of the vocal organs, facial muscles, arms and hands expire, so further in terms of our conditions it is conceivable, that the sound waves caused by the muscle contractions can be thought of as a new complimentary condition Rε which causes a change of the partial system c_3, from which the change in turn transfers to the first-changed partial system c_1 to a certain extent. The same change circle exists in the system of a second individual, that is the motion Rε may originate at C_2 as it is with the first individual from system C, and vice versa if the circle starts from C_1, then at C_2 the same changes exist as in the form of c_1 caused by Rx: the movement or the sound thus receives the meaning of a complementary condition Rε, which is able to represent the original complementary condition Rx (a complimentary of R).

126. — Assuming that a majority of propositions of the same complementary condition Rx, or of more or less closely related ones, have entailed a series of more or less related changes of the partial system c_1, each of which again expresses the representative complementary condition Rε in the manner just given; Therefore, according to our general presupposition, it is conceivable that on the occasion of the assertion of Rε, there will not be a single change of c_1, but likewise any (to be determined in each case) majority of related changes take place in immediate succession.

Second Section

The Preservation of the Individual

Chapter One
General

I.

127. — Our decomposition of the changes in C has, to this point, retained the relation to R insofar as either R should have the meaning of a *complimentary condition* for C, or C should be given the meaning of a *systematic precondition*.

Now, however, the notion of a systematic meaning of central partial systems, obtained in (no. 123), leads us to a further differentiation of the changes of C, which likewise finds its presupposition in the context of our general assumption

It was there in (no. 37) that the human individual was presupposed as one who asserted himself, at least within certain limits, by reduction (*Verminderungen*) for his preservation.

Based on this concept of presupposed assertion (*vorausgesetzter Behauptung*), we can give the viewpoint that gave us the meaning of the partial systems for system C thus far.

Based on this notion of presumed assertion, we can generalize and augment the point of view that gave us the meaning of the partial systems for the system C , as far as we first consider the notion of a meaning of the system C as the assertion of the total system to which it belongs as an organ, that is, by adding to the whole organism, and then to those of importance to other systems to which C also belongs to by virtue of the fact that its changes add to these changes and may themselves be subject to change.

128. — The new way that opens here, and leads to a distinction of the changes of the system C according to their significance for the assertion

firstly of themselves, then of those particular systems of higher and highest order, in which the individual human is classified – this means an understanding of system C in its relation on the one hand to the family of systems, of the state, of the church, of society as a whole, and on the other hand to the systems of the nearest and wider environment of the earth - the system R, or as individuals say 'the world' overall.

In the matter-of-fact-way - from the simple partial systems of System C - and finally to determine the relation of man to society and the 'world' both theoretically and practically – on this path into more distant and sublime spheres, we now stop here and try, now to see the importance of system C for the preservation of the organism in general – and sketch with a few strokes, sufficient enough for our next goal.

129. — The meaning of system C for the preservation of the entire organism, whose subsystem it contains, results from its concept as a central system which collects changes emanating from the periphery and distributes the changes to the periphery (see no. 71 and before).

Thus the more peripheral parts of the organism, the greater the importance of C for the maintenance of the organism, which is more directly exposed to the environment and therefore, on the one hand, suffers the changes set by each individual R first, however, on the other hand, the totality of R can turn out to be regarded as a change condition (see no. 85) – the richer and more varied these peripheral parts are represented in C, the more intimately consequential the preservation of the organism C in its functional and formal design.

130. — The more intimately, after, and the higher C is developed, — the preservation of the organ must then be bound to C, but greater meaning must be attached, for the preservation of the organism, to the preservation of the system C itself.

Most (or all?) Consider only the maintenance of the organism by the system C, not the preservation of the system C itself.

II.

131. With regard to the preservation of system C we can now consider the following cases.

1) There are no change conditions set;

2) Change conditions are set, but the formal nature of the system does not permit its destruction;

3) There are set change conditions, and the system's properties would cause destruction by those changes, but the system maintains itself under these threats (*Bedrohungen*) of its formal existence from any further changes to itself.

In case 1, conservation would be simply an inertia, in case 2 a formal indestructibility, in case 3 a vital conservation.

In the latter case, we should designate the variable size of this vital conservation as the vital conservation value (*Erhaltungswert*) which would be attributable to the system C at each point in time of its existence.

In the latter case we should designate the variable size of this vital conservation as the vital conservation value which would be attributable to the system C at each point in time of its existence.

Case 1 would be equivalent to the case that C would be environmentally non-existent (see no. 59) Case 2 and 3 allow the assumption of an environment, but only in case 3 is there the assumption of such an environment, which can become threatening for C.

132. — Which of these cases we consider for our purposes depends on their compatibility with our general assumption. Neither the concept of an environment-less C nor the notion of a formally non-threatening environment is compatible with our general premises; and this appears because the third case — the formal threat — does not contain a contradiction. So, we would first have to reflect on the importance of the environment for the vital conservation of the system C.

133. — We now make a fiction that seems methodologically necessary, as it is certainly methodologically permitted. We set up a fictitious environment that in no way permits a reduction in the vital conservation value C. Thus, this environment does not contain any moments, which would be detrimental to the vital preservation of C, and so we may call such an environment *ideal*.

Any other environment would then depart from the ideal one all the more, so the less the conditions of the ideal were fulfilled – meaning the more conditional moments contained, the more reduction in the vital conservation value of C.

If, as we have said, this *ideal environment* is a fiction, then these environments, which are distant from the ideal, are justifiable assumptions (see above, no.132)

134. As little as we think of the environments in which human beings suffer and struggle in from birth as *ideals*, the environment of man cannot be denied an approximation of the ideal conditions. This is not in conflict with our general requirements. For this is the only environment in which man has not been displaced by his birth: it is the environment in which he was before his birth - the maternal nourishing and protective shell.

The "suffering and struggling people" were said, and what kind of man do we presume other than suffering and struggling, though not exclusively as suffering and struggling.

135. — From this womb, this sanctuary of preservation, the child is expelled; thrust into an almost completely different, new, unfamiliar, an only partly still preservation-friendly environment. Now the child is exposed to the changes that arise from the environment and its transformations; and soon the human will be exposed to is destinies, which impose on it the typical changes of a particular course of development.

And this means: The system C has been transferred by birth from an approximately ideal environment to a non-ideal environment.

136. — Here again we have come to a new point of view from our general empiriocritical presupposition which is a distinction in the changes of the system: namely in their relationship to the vital conservation value (*vitalen*

Erhaltungswert) attributed to system C at any given time. And our task will therefore initially be directed to analyzing the changes in the system C: if they be thought of as *diminishing* the vital conservation value of system C, or as *assertions* of that system under such dimunitions.

Second Chapter

The Vital Conservation Maximum.

I.

137. — At any point in time t, let us put down any reduction in the vital conservation value of system C to be assumed at that time, so there are always 2 conservation values (*Erhaltungswerte*), w_1 and w_2, thus w_2 which results from the reduction, and w_1, which was reduced which can also be thought of as $w_1 > w_2$.

If I now think of the difference between w_1 and w_2, then I have only thought of the value by which the conservation value of C was reduced, especially at time t. Now suppose, at the same time, that the environment in which the system C was transposed at birth to be non-ideal, the less an approximation from the ideal, the less is the immediate value of w_1, which was diminished and immediately accepted as an *absolute*. To accept one without further ado would be permissible only in the one case, which has been excluded, namely in the case of an ideal environment. Thus, in the relationship $w_1 > w_2$, initially only one reduction difference is expressed: but by itself, it says nothing about how much the vital maintenance value is diminished or how great the whole diminution has become at all, when the vital conservation value sank from w_1 and w_2 at time t.

138. — So, should I be able to think of the whole diminution of the conservation value of C, and thus of the value of vital conservation set at

each point in time, then I must be able to relate any conservation value, which is considered to be diminished, to a conservation value that is no longer conceivable as *diminished*. As far as I can think, in only one case is a conservation not conceivable as *diminished*. This vital preservation value of system C is required for the determination of all reductions and therefore must be conceived as such.

II.

139. — Since, like the whole system C, its components e.g., their form elements are created and pass, for some point in time, the requisite maximum vital conservation value of system C is assured, then it must be used to maintain the absolute preservation of all central partial systems e.g., including form elements; thus, conceivably the greatest vital conservation value of system C is the sum of the greatest conceivable preservation of all its components e.g., form elements.

140. — The formation of C is not isolated here, but connected with the entire organism and as it exists, it is generally thought of as expiring with it.

Even a relatively isolated demise of C may not be unthinkable, in any case, as is any individual, relatively independent partial system of lower order to be included in our general supposition as with parts of physiology (and psychology).

141. — We reflect on the isolated demise of the simplest partial systems e.g., form elements, emphasizing first of all, that if the individual components can be considered subjected to degradation, their preservation may not necessarily be thought of as unconditional. Rather, an existing partial system must have conditions of its preservation — passing changes are considered in its conservation condition.

The preservation conditions have the priority, because the partial system was assumed to be "existing".

142. — As a fundamental conservation condition (*Erhaltungsbedingung*) next the task would be to think again if at the same time R is thought of as the practitioner; for to the extent that a partial system lacks all the changes

(= activities) that are set with R, the opposite of the sequence of activities would also exist (compare no. 117) — namely – instead of positive – negative increases and forms, thus a change would be expected, which should be described as degeneration and as an increasing approach to deterioration (Untergang).

We have already labelled the system changes with R set as f(R) (no. 89)

143. — At no time is there a complete cancellation of R in its entirety, that means the entire environment, or a complete exclusion of the changes emanating from it, also the lack of f(R) is never to be considered as an absolute, but in all cases only relatively. Consequently, the presence of f(R) is considered as continuous.

III.

144. — But if the relative lack of f(R) is to be considered as a condition of degeneration, then besides f(R) other changes of C or its form elements must be assumed, indeed since a *lack* of oneself cannot cause changes, other systematic changes as a *consequence* of degeneration means to think of the demise of the form elements which could be thought of as the annhilistic conditions of (system) C.

145. — However, changes in C must still be considered differently than those accessible with R, just as changes f(R) cannot be thought of as the only kind of changes in C. This second type of change, whose C value must be considered (if its components would approach degeneration due to lack of f(R)), cannot be thought of as originating from the environment R in the same sense as f(R); because if they were simply dependent on R as they are on f(R) we would have to again think of f(R) as not having different kinds of changes.

146. — Since C can only represent the totality of the systemic preconditions for this other type of change, a change condition not included in the change in question — as a complimentary condition — is required for the setting of changes in question; but since the environment can only be an altered condition for C in the sense of R or S, and since R has finally dropped out of

these two conditions of change in the present case, S is to be thought of as a condition of change.

Accordingly, we designate this second mode of change, in agreement with no. 89, again as f(S).

147. — Also, these changes in f(S) are considered continuous and in existing since neither the circulatory system through which the environment components of type S are supplied to system C and its form elements nor the process described by physiology as metabolism can be thought to be completely absent at any time unless death has been set.

IV.

148. — If we consider the preparation and circulation of blood as well as metabolism as a constant quantity, then we also think of any corresponding changes as f(S). Similarly, the results would approach the end if there was a sufficiently long period lacking of f(R). It would move away from this extreme if this deficiency were lifted in time. From this dependency of the change direction of f(R), it is clear that f(R) and f(S) must be thought of as opposite changes.

149. — Considering the same law with which f(R) could previously be described as a preservation condition and f(S) as an annulment condition, conversely f(S) can now also be called a preservation condition and f(R) an annulment condition. If one always assumes the value f(R) is always decreasing by f(S), then it is just as important to presuppose a corresponding approximation to the decline (degeneration) of C or the effected form elements.

150. —But if we believe that there is not only an approximation to decline (degeneration) in the mass, f(R) being *smaller*, but also in the mass, when it is greater than f(S); consider that within the mass itself if f(R) is smaller or larger than f(S) then the reverse is that f(S) is greater or less than f(R): thus, we can no longer address f(R) or f(S) *per se* as conservation or annihilistic conditions.

151. — Rather, it is the **difference** between the two types of change f(R) and f (S) in the mass that designate the annihilistic condition, as both depart from equality; and then as the conservation condition, as both values approach equality.

V.

152. — Therefore, the vital conservation of each form element will be complete if the following equation holds for it:

1) $f(R) = - f(S)$.

153. — And thus, the vital conservation of the central partial systems, which form the system C, and thus constitute the system C itself, will be complete, if the following equations apply to the central partial systems:

2) $f(R_1) = - f(S_1)$

$f(R_2) = - f(S_2)$

.

.

.

$f (R_n) = - f (S_n)$,

Consequently, for the entire system:

$\Sigma f(R) = - \Sigma f(S)$

154. — But since (according to no. 148) f(R) and f(S) are opposite values, the complete vital conservation of the formal element can also be expressed by the equation:

3) $f(R) + f(S) = 0$.

155. — Accordingly, we get the equations for the central partial systems:

4) $f(R_1) + f(S_1) = 0$

$$f(R_2) + f(S_2) = 0$$

.

.

.

$$f(R_n) + f(S_n) = 0,$$

consequently, for the whole system C:

$$\Sigma f(R) + \Sigma f(S) = 0.$$

156. — We assign the values $f(R)$ and $f(S)$ as they relate to the partial system as partial systemic factors, and their sum in relationship to the vital maintenance value as the vital difference, so that it follows that if the complete vital preservation depends on the opposite equality of the partial systematic factors, any vital conservation value thought to be smaller than the complete total must be dependent on the vital difference.

157. — The preservation of system C is complete when the vital difference is zero, since a vital difference cannot be thought of as less than zero — and no vital conservation value can be considered greater than that which the system possesses if the partial systematic factors are equal. It is therefore the vital maintenance value which is set by the vital difference to zero, its greatest possible value.

158. — In denoting the greatest possible conservation value of system C as the vital conservation maximum, we obtain the proposition:

A system C is to be considered at its vital conservation maximum (*Erhaltungsmaximum*), if its partial systematic factors are equal. (*see footnote no. 5*)

Third Chapter

The Fluctuation.

I.

159. — The establishment of the vital conservation maximum leads to a distinction of a changelessness of C, which must be set if no change condition is set and a changeless state is arrived at when a majority of opposite changes are adopted. The former is considered notions (*ex notione*), the latter forms (*ex specie*) which means it appears far from being free of change, rather their setting appears to require a double change of form. The first type of changelessness may be distinguished by the term systematic persistence (*Systembeharrung*) and the second as the system's resting point.

160. — Changes occurring when the system is at rest are referred to as system fluctuations and are positive fluctuations if the changes to the system are through positive propagations (*Vermehrung*) — as negative if the change is due to a negative in one of two partial systemic factors.

161. — On each (positive or negative) fluctuation we think of the fluctuation as being complete (completely expired) if the system resting point has been re-established; and accordingly, each time a fluctuation is initiated, if it is thought to be completed, there is an increase and decrease in the change of the original system state.

This provides the distinction of a positive or negative fluctuation for the general fluctuation.

162. — The value by which the resting system C is increased when a change condition is set, that is, the difference from the system at rest, is called the fluctuation quantity.

163. — For these fluctuations, it is first necessary to claim the state which the analysis revealed for the changes in the central partial systems in general (no. 118): the differences of form; thus, the form of fluctuation, which (according to no. 124) is conditioned not only in the specifically complimentary condition, but also in the specially determined preparation of the system C (e.g., of the central partial system).

164. — If the developing condition (*Entwicklungsbedingung*) (reference no. 119) allows more or less related complimentary conditions for the partial systems, this also results in system C variations, i.e., its partial systems, form or related form characteristic, whose upper limit values are referred to as relatedness in the narrower sense, and whose lower limit value was designated as opposite in the narrower sense (no. 120).

165. — We still get the concept of fluctuation relevance from the combination of variable change and the systematic expression of the changed partial system - this means we perceive the fluctuations as a moving contrast of relevance and irrelevance, as do the partial systems (see no. 123), and the fluctuation of a central partial system is all the more relevant as changes occur with systematic significance.

166. — From the differences of the fluctuation on the vital conservation maximum, as those are set to positive and negative, we gain the concept of fluctuation direction and designate it in the first case as positive, and in the second case as negative.
no. 160 - 162 the fluctuations were in relation to the system rest (Systemruhe), this means changelessness by mutual cancellation of two changes; here the relationship goes to the maximum conservation that is set with that changelessness. These two relationships correspond to the two sides of the object: the purely physical (quantitative) and the biological (qualitative).

167. — Let us call a system change which directly captures a certain peripherally set change condition in a partial system, a *primary* one; and another which in the propagation of additional partial systems, joins the primary one which is called a secondary one: so with respect to the *secondary* change, one may assume that it either remains within the sensory partial systems or the motor e.g., secretory partial systems - according to the nature of "reflexes" in the physiological sense - overlaps (see no. 114). At this

point, the variations of system C, set in the former manner, will offer no further peculiarities requiring special attention. Conversely, the fluctuations of system C, which are based on a "reflexive" overlap of systemic change on motor or secretorial partial systems require special mention: providing for fluctuations of the motor e.g., secretory functions associated with sensory partial systems (see no. 78). Therefore, we want to call the latter type of variation, as opposed to the more limited former as overarching fluctuations.

II.

168. — Further, general determinations of fluctuations are obtained if we reflect on the activity whose systematic expression was the formal and functional determinant of the partial system, and by which, the form of change was co-determined.

If we assume a fluctuation in order to fix these values, which have remained complete in the sense of the previous activity of the partial system, we have assumed an exercised fluctuation. If we then subject this *practiced variation* to a change, we obtain the presupposition of a fluctuation variation — under which we always seek to understand the variation of an expressed fluctuation.

169. — With the fluctuation variation as a change of a practiced fluctuation - expressed otherwise: as a deviation from certain activity direction of the system C — consequently a change of all circumstances coincides, which are themselves based on exercise; for if, according to the assumption, the practiced fluctuation was an entire expression of the preceding activity of the associated partial system, then the variation will be less than the original *change* as it represents just a change result.

170. — The variation is now affected by:

1) the form of the fluctuation — this is due to the formal and functional certainty of the partial system, this is in turn through the activity.

2) the expressed value of the fluctuation — because it can be set to x times, but also xy times, and in the second case set, (if set to y), a different value can be obtained.

3) the totality of connections (*Zusammenhänge*) in which there is an individual nature of the variation with other qualities, or an individual variation with other variations (according to the concept of the system).

171. — If I express a sustained fluctuation, on the one hand I remove the changes from system C from its established form; on the other hand, I set a system change, which at the same time has another, initially at least lower expressed value: if he expressed fluctuation as such was also the change, then the varied fluctuation is the less experienced.

These removals of a change of system C from a trained form is considered positive — the conceivable reverse case, however refers to the approximation of a system change to a practiced form as a negative (*Transexerzition*); "for the practiced form - here is the practiced form approximating the system change which need not be the same from changes previously removed."

172. — We call the particular activity value the respective fluctuation expression or the expressed activity.

173. — If we speak of the practice of fluctuation - of the exercising thereof - we then understand both together.

What of the form of the practiced fluctuation, e.g., their variation has been mentioned here, which also applies to their *size*: this too can be practiced, the exercise thereof formally and functionally co-determining the associated partial system — their *variation* also goes along with the change in the total expressed values.

174. — If then the practiced fluctuation was not absolutely isolated, but always expressed at the same time in certain internal and external contexts, these fluctuations are also brought about by the fluctuation variation from the stable uniformity of their mutual relations: the fluctuation changes from relative uniformity of connections to a greater variety in fluctuation variation: the fluctuations become more agile, differentiated, articulated. At first these expressions capture only the opposite of the preceding ones. It is

permissible to refer to the transition from an (unvaried) practiced fluctuation to a varied one according to the observed fluctuation articulation.

175. — If the characteristics at each timepoint determine the meaning of the fluctuation, the characteristics in the fluctuation variation move toward or in a narrower sense in contrast to relevance and irrelevance of the stationary partial system as well as toward a positive or negative direction, and further toward a positive or negative transformative exertion, the fluctuation activity increasing or decreasing the expressed value — I say that the varying variations include features moving in opposition, and it follows that even the variations of variety (*Schwankungsvariationen*) can maintain the sense of the original fluctuation or assume an opposite meaning: in the latter special case we want to designate the expressed fluctuation more closely related than fluctuation opposition.

176. — The transition from a practiced fluctuation to a varied one is all the greater, the greater the extent of the change and the faster it is set.

III.

177. — Finally considering the relation of the oscillation variation to the vital preservation value yields differences of fluctuation which we regard as ordered.

Every practiced variation, insofar as it is believed pure as such, that is without variation, should simply be referred to as a variation of the first order.

Each change in such a variation, as far as (the change) provided at the same time diminishing the vital preservation value attained at their entry, as variation of the second, third etc. order.

Following this, these changes which are made to the equality of the partial systematic factors (no. 159) - in cases where *the fluctuation* has assumed the value of zero - it may be added to the variation of order.

Although every fluctuation of a higher order is the variation of another fluctuation, not every variation of a fluctuation is a *variation of a higher order*. It may be the decline of a variation to a lower order; or a first-order fluctuation.

Fourth Chapter

The independent vital series.

I.

178. — Since our first task was to analyze (see no. 136) the changes in system C, as far as they are thought of as diminutions of the vital conservation value of system C or as assertions of system C under such diminutions, we formulate the result of our previous analysis firstly for a general determination of diminution as follows:

If system C is to be considered as reducing its vital maintenance value, it must be provided with a vital difference greater than zero, e.g., this difference must be conceived as a fluctuation expressing a positive increase.

179. — And therefore for general purposes the following statement follows the sentence:

If a system C is to be thought of as being subject to diminutions of its vital conservation value, its assertion must be approximated to zero as the approximating the difference in vitality set in the reduction, meaning it be thought of as negatively increasing fluctuation.

180. — If, therefore, it is assumed that the system C completely asserts itself under any reduction of its vital conservation value, then at the same time a positive and a negative increasing fluctuation, which correspond, are assumed.

181. — If, however, it is further assumed that the system C has completely asserted itself or will assert itself under any reduction of its vital

conservation value, it would posit that the environment by itself does not (or only irrelevantly) condition the negatively increasing fluctuation, and it is at the same time assumed that the system C itself will pass or make changes which will lead, directly or indirectly to the relevant increasing negative variability.

Every assumed series of changes which means an assertion with diminution of the vital preservation value of C, we briefly call an independent vital series.

In the following, the term vital series simply means: to be understood as an independent vital series unless explicitly noted.

II.

182. — Assuming an independent vital series as complete (completely expressed), it follows from this concept:

Each vital series, if complete, can be thought of as being divisible into three parts or sections, the first of which introduces the positively increasing variation; the third section forms the completed state while the second includes all changes placed between the first and the last parts.

We call the first section the initial section, the second is called the medial section and the third is the final section; and the associated changes or terminations are referred to as the initial, medial and final changes or terminations.

The apparent contradiction between the timelessness of the "perfect suspension" and the duration of the final section is resolved by the fact that either an after-time duration analogous to the duration of the optic stimulation or sensation of movement, or an oscillation takes place.

183. — If the concept of the case analyzed shows that a complete series of vital signs must be composed of at least three sections under the given conditions, the description (no.181) shows that such a series is at the same time composed of three types of changes.

184. — But neither from the concept of the assumed case nor from the concept of the vital series and their types of change does it follow that *every* type of change in the series, and therefore each of its sections, must be represented only by one section each, that is the vital series, although consisting on only 3 sections must be composed of three sections.

On the contrary, if it is conceivable that the diminutions of the vital conservation value, as well as the claims of the system C, may be placed under such reductions in the form of successive changes, it is also conceivable that one or more sections of the independent vital series can consist of more than just one of its sections. (see The Graphic Presentation of the Fluctuation of System C)

III.

185. — In considering the immense variety of conceivable environmental changes, but also in the systemic conditions on which central partial systems depend – these must be finite as are the majority of changes within each type of change (*Änderungstypus*), so that the course of an assertion of the value of vital conservation, as it then may be assumed for the system, can be thought of as being composed of the majority, be it adjoining one another in many ways e.g., consuming vital series, so that the analysis of the changes in which system C is asserted in a non-ideal environment is not just a single tripartite series of changes, but arbitrarily many-units, even whole systems of vital series and systems of vital-series systems comprised of relatively simple to the most complicated compositions — and all this even if limited to a few special cases.

The case that the assertion of system C follows e.g., it will occur if the negatively increasing fluctuation in the one main partial system occurs at the same time a diminution of the vital conservation value is set.

186. — It is not our task of our general analysis to trace in detail the successions, intersections, and entanglements that can be formed by vital series and vital series systems; We will probably have another opportunity

to look back on these bands, nets and — bundles. At this point, we have to humble ourselves once again with simpler pieces of the immense substance: and to look for those general features of the (complete) vital series, which for us are the more important ones.

187. — For simplicity sake, let's start by looking at relative requirements of simpler work, such as a result on system C of secondary changes of any vital differences in a non-ideal environment that would be cancelled out in a simple and sustainable way, thus we would have, as before, an "ideal environment" or an "ideal system C".

As mentioned earlier, the ideal environment and an ideal system C is treated as fiction; Justification may be claimed only for the assumption that the system C approximates the ideal of the various organisms in very different forms and stages — according to the species to which the organism belongs, and according to the degree of its general as well as individual development e.g., degeneration. Any general description of the vital series, therefore, could apply to the individual only with the limitations imposed by the genus, species, and subject matter of a particular case.

IV.

188. — From the concept of a completed vital series it follows further:

If a complete vital series is to be thought of, it cannot be thought of as ending until a change is made, with which the vital difference is thought to be cancelled.

189. — And finally:

If a completed vital series is to be thought of, it cannot be thought of more than the change with which the suppression of the vital difference was made. Because with this change either the conservation maximum is reached, and then every further change must reduce the maximum value of the preservation, thus setting a new vital difference and consequently be thought of as part of a new vital series; or with any change made at a certain time, the maximum conservation value is not thought to be met, and then the condition under which a vital series can be considered complete is *not* fulfilled.

The Initial Section

the independent vital series.

First Chapter

The vital differences in general.

I.

190. — An important distinction in the vital differences, whose significance is unlikely to come to light in the following investigation, is only mentioned here: the difference between the general and special vital differences. It is conceivable that the variation is limited to one or more specific partial-systems; But, it is also conceivable that they are pathological in general nutritional disorders; - effecting even the whole system C. Our general theory will have to deal with only the fluctuations of special partial systems, thus with special vital differences. These are therefore, unless otherwise expressly noted, hereinafter applicable everywhere.

II.

191. — If we now turn our attention to the conceivable cases of the positively increasing fluctuation, their nature and classification must be captured by the varying of their formal expression.

Since the partial systemic factors of the same are opposite in the vital conservation maximum of any central partial system, the equation holds for the case where the vital difference is equal to zero.

$$\partial == f(R) + f(S) == 0.$$

All conceivable cases of such variations of the partial systematic factors ∂ in which the equation:

$$\partial == f(R) + f(S) == o$$

Inequality is thus expressed

$$\partial == f(R) + f(S) > o$$

Consequently, all the cases of conceivable positive increasing fluctuations result; and conversely, all conceivable cases in which the inequality yields

$$\partial == f(R) + f(S) > o$$

The equality is thus expressed

$$\partial == f(R) + f(S) == o$$

And the same is attributed to all conceivable cases of negative increasing fluctuations.

192. — The simple forms of the positively increasing fluctuation now result from variation of one of the two factors $f(R)$ and $f(S)$ in the equation. The variation of the factor $f(R)$ can then be referred to as the working fluctuation and that of the factor $f(S)$ as the nutritional fluctuation; The cancellation of the system rest which is based on a positive increase of the partial-system factor in question, has already been distinguished (no. 160) as positive, and the negative increase as negative fluctuation.

193. — Each of these four simple types of positive increasing variation, obtained by positive or negative multiplication of one of the two partial systematic factors, then again allows two simple types of negatively increasing variation:

1) The varied value itself can again be increased by a change of opposite sign;

2) The opposite value can be increased by a change with the same sign.

These simple types of positive and negative increasing fluctuation could be understood as the basic types, through which the further variation of all other conceivable forms could be found and divided.

194. — In the forms of diminution of the vital difference, as well as in all types of changes which cause diminished forms, these contain the formal

conditions which a system change must then fulfill to release a vital difference.

III.

195. — The development of all conceivable forms of vital differences and their annulment according to (no. 191) means this whole method of finding and classifying all conceivable cases of positive and negative increasing fluctuation is only likely to become fruitful if it were applied on a scientific basis which is yet to be attained.

But if we have to content ourselves with a more modest procedure in the meantime, it does not follow that the general idea of the suggested classification method should remain entirely unused; but the selection of conceivable cases, which is expedient in the sense of our task, must seem more urgent to us than a complete enumeration, which is of course desirable in itself.

196. — Therefore, we distinguish between significant and insignificant differences in vitality. We refer to the fluctuations that are so minor that they do not exert any reasonable lasting difference on system behavior of a system C as insignificant differences in vitality. All other vital differences on the other hand, which are thought to exert a further influence on the behavior of the system for our investigation, we also call significant differences in vitality.

Our task does not require dividing up the conceivable significant differences in vitality; We hope to satisfy our purpose if we record and label only those significant differences in vitality which we are likely to encounter.

Second Chapter

The Specialized Cases.

I.

197. — In order to be able to proceed sufficiently simply also in the instruction and arrangement of the vital differences to be treated, we first select a case in order to base it on the determination of all other cases - in such a way that we take all other vital differences into account as modifications of the vital differences of the selected case.

In this choice as we will briefly say of this fundamental case, we have to decide whether to select a fluctuation of a major or minor partial system, an expenditure or nutritional fluctuation, a positive or a negative increasing fluctuation, or one within or outside the physiological conditions being activated as uniform or unevenly set (at least approximately), and in turn a simpler or more complicated process; of an individual still in the process of progressive development or one that is already expressed.

Again, the aspects of simplicity and fruitfulness are decisive in the decision.

198. — In the former sense, the result appears to be from the uniformly conditioned nutritional fluctuations, which is simpler than that of the many-sided conditional fluctuations of work expenditure; The results output of those things placed in conditions of rest are simpler than those set in the colorful game of awoken life (*im bunten Spiel wachen Lebens*). The result of the positive side of the positive dietary fluctuations are easier than the negative and the negative dietary fluctuations, the latter leading to the a study of system C behavior as food related rather than our task which focuses on forms of expenditure (*Arbeitsformen*); Moreover, the result is easier to understand from the (almost) uniformly nullified fluctuations than from the irregularly set and unevenly nullified ones; And finally, the outcome of major partial systems, which already reoccur functionally and

formally, appear simpler than sub-partials which would first have to be developed.

199. — We are guided in terms of fruitfulness of a given case by the following reminder:

In the womb, as an environment approximately constant with respect to R values, the value of f(R) is also considered to be comparatively constant, while the value of f(S) is bound to changes in maternal food intake and thought to be subject to many variations. In any case, these dietary fluctuations of the unborn child cannot be linked with our special E-values, as long as the system C is thought to be placed under the conditions of rest before birth. The child's dietary fluctuations are therefore connected as special E-values, then if the child has been ejected from the maternal lap – by the violent act of birth- it is from then onward it is sustained by its own daily nourishment rhythm and system C is temporarily relieved of the conditions of rest.

But when contemplating the individual whose reoccurring central partial systems have been functionally and formally determined appears simpler; And so, the progress of our investigation must seem more fruitful to us if the individual forms the fundamental case is neither before or immediately after birth, but already in a phase of life in which it has reached developmental stasis (*Entwicklungsstillstand*).

200. — As a reminder, we favor then a positive and positively increasing rate of change for the (selected) basic case, which is set in conditions of rest, but is unset within the conditions of being awake; We prefer a positive and positive increase in dietary variation, because our objective is to study individual stages of culture or forms, which the search and intake of food is not the most prevalent conservation condition; it must appear more productive than the study of cultural levels or forms whose work arises from the purchase of food; we choose a uniform work increase offset with usual dietary fluctuation because the more an organism is trained, the more regular the condition of conservation. We select the main partial-system because we have to ensure the relevant fluctuations, and indeed we choose

those partial systems which are made up of secondary partial systems — of approximately uniform reoccurrence in a given rhythm of nutritional and labor expenditure changes and depending on their sizes and forms — developed into functionally and formally determined main partial-systems. And all this in an individual who is still in a process of progressive development, because only the investigation of such individuals can promise to become of practical significance at the same time.

Considering the latter (formal main partial-systems), it does not need to be ruled out that this development can be a priori supported by an inherited or otherwise innate system.

201. — We then choose the case to which we base our investigation:

a central partial system, which belongs to a vital individual, being provided with an equal nutritional increase (based on the systemic unit or earlier acquisition) when the conditions of waking are set: think of an increase in work, which - with the conditions of being awake and with size and form also equally set - the nutritional supplement being sufficiently uniform and sufficient for a long period of time - in order to develop (*zu entwickeln*) the associated partial system into a differentiated function and form of the main partial system through sustained and directed practice.

The *uniformity* should only be claimed for temporally close moments of development - the emergence of larger differences with moments of development that are more distant in time, therefore, should not be allowed to be taken into account.

II.

202. — We refer to the uniform nutritional increase set in the selected case, with which a main partial system is thought to be provided when entering the conditions of waking, from the partial systematic co-moment Π; The corresponding uniform increase in labor expenditure, which – assuming the selected case - has been annulled sufficiently regularly and for a long time each day, is considered as the partial systematic co-moment Γ. And accordingly, we shall call the elevation of a change of labor to the value of a partial-systematic co-moment as positive, and the reduction of this value as a negative co-moment.

Not every fluctuation of the system C means a partial systematic co-moment Γ; but every partial systematic Γ also means a system variation.

203. — The settlements of the partial-systematic moments Π are now provided for by the division of the vital differences to be treated by us, as we further define it, following our distinction of the orders of variation (no. 177) and the distinction we designate as vital differences of the first order.

204. — All other possible cases of vital differences would be obtained by variations described in no. 201; therefore, the most general and simple modifications, if one varies only one or two values in Π and Γ and leaves the remainder constant.

In each of these cases, the vital difference cancellation, which was set by the combination of the unchanged values, is subjected to a modification — by thinking of both cases, therefore, there are two types of easily derived vital differences. We call such derived vital differences according to their elaborated settlement as vital differences of the second, third, etc., order.

205. — Such vital differences of higher order can be achieved in the case of the factors Γ both by their quantitative and their qualitative variation; and from the momentary factor Π by multiplication or diminution: this then from the previously uniformly established nutritional increase.

206. — Finally, a special case of a change of the co-moments is conceivable, which, because of its importance for our purposes, deserves special mention.

We gained a second-order vital difference from the factors by thinking that the environment of their setting was different; it is also possible to vary the co-moments independently of the environment in a certain sense — namely, by coexistence with a second partial-systematic co-moment, which is thought to be a change condition in its relation to the former.

207. — The change of factors due to qualitative deviations requires a brief remark. — The fact that the equality of the partial systemic factors is abolished, even though the set work increase in production has remained qualitatively the same as that of the original, but deviates quantitatively from

it, requires no further explanation; If the set increase in labor remained quantitatively the same and deviates only qualitatively from the co-moment (*Komoment*) it follows that the qualitative deviation can also be reduced to a quantitative expression: According to our assumption, the changing partial system in its form of change has been determined functionally and formally by the approximately uniform and sufficiently long acting environmental constituents; If we think of the partial system as a certain form of change suitable for the co-moment, we at the same time think of a restriction or abolition, and consequently of a change in the (formal and functional) determinations which have been developed for the partial system, upon which the co-moment rests.

208. — Finally, the vital difference of the value zero, as it is set in the vital preservation maximum, is called, according to the corresponding fluctuation description (no. 177), a value of the first order.

III.

209. —The indicated vital differences are introduced as conceivable, as such, they do not contradict the concept of system C. We take another look at their relationship to our general empiriocritical premise.

As far as the selected case is concerned, our assumptions are likely to remain the same; in particular the setting of uniform work fluctuations e.g., sufficiently recurrent environmental constituents into which an individual is first put from the day of birth, and then again daily (see no. 22) and the development of the partial-systems as a result of that exercise which formally and functionally determines the particular partial systems in the results of our analysis (see nos. 76 and 118). Since, however the uniform work activity must go hand in hand with a fortifying nutritional habituation, otherwise the uniform activity could not be set in the long run, so the habituation to eating can offset the increase in work as well as be annulled by the increase of expenditure from work, so too, it is the condition for accepting and applying the selected case to a uniform increase of work, e.g., contained in relation to the uniform nutritional increase in our general presupposition.

210. — Also the vital difference of higher order is included in our premise.

Since the selected case is based on assumption of a uniform rhythm of nutritional and work fluctuations, meaning the regular return of certain dietary and work increases in a formally and functionally determined partial system; moreover the determination and definiteness of all these values have their own conditions in the nature of the environment and of system C itself; and these conditions were finally assumed by us to be changeable (see no. 22): Thus, with the permissible changes of conditions, we must also allow the changes of the conditional, hence such changes, be they in the surrounding components or in system C itself, with which a deviation from the previous regularity (or uniformity) is considered with respect to the co-moment Γ or with respect to the moment Π itself.

211. — In particular, in relation to the environment, it follows in all cases where the co-moment Γ (labor expenditure) – in a degree – depends on the magnitude of the condition of change of R, or is specifically intended as a mathematical function of time or distance, that quantity or time or distance in which R is set — or, in all cases where the co-moment Γ - according to its form - is thought to be dependent on a particular species or combination of environmental constituents, that species or combination – which is given to be changed by any alteration that may be added, and thereby constitute the condition for the setting of vital differences of higher order.

212. — On the one hand, it fulfills our requirements, after all the qualitative and quantitative relationships in which a nutritional variation is thought to be dependent on the supply of nutrients, the inhaled air, the air pressure, the temperature, the physical movement, etc., that these conditions, individually or together, seem to be varied so that in the first place the setting of the hitherto uniform dietary variation is not permitted in this case.

213. — On the other hand, it is also in accordance with the results of our analysis, if we, by maintaining the external conditions of nutrition, be it through progressively augmented or diminished formal transformations, or by one of the typical physiologically developmental processes, or by unfolding a pathological system (inheritance), also consider the nutritional certainty of a central partial system and thus the associated change moment Π.

214. Finally, the assumption of possible differences in vitality of the special case noted (see no. 206) is included in our general condition, provided that system C can be put in place by concurrent conditions of change consisting

of several relatively independent practiced fluctuations, whose form is related e.g., whose forms are partly identical, partly different, and each assumes the meaning of one condition of change for the other and are now at the same time two cumulative co-moment fluctuations and such changes would give rise to a more complicated special case of vital differences of higher order, which we have pointed out. Nevertheless, the distinction of orders - here as well as elsewhere - is not so keen or as rigidly structured for practical reasons as might perhaps be desirable in itself. Our orders are only intended to give a very simplified and schematized, but also an accessible, picture of unlimited gradations, shades and outlines.

IV.

215. — Suppose that a partial system, as assumed in this selected case, is provided with an arbitrary partial systemic moment at the time it is put under the conditions of awakening with partial systematic moment Π, — it must be assumed that system C is subject to diminutions of its vital conservation value — so that a corresponding negative increasing fluctuation is required.

If the moment Π (nutrition) has been set, then the conditions of its cancellation, e.g., the conditions for the setting of the associated co-moment Γ is presumed,

Case I: unchanged,
Case II: changed,

216. — Case I. If the unchanged conditions are presupposed, then the vital series can again be thought of in the same way it was completed. We want to describe the simplest case of a series of vital events – which in our case, does not offer any special multiplicities - as a first order vital series (*Vitalreihe erster Ordnung*).

217. — Case II. If changed conditions are presupposed, they can be presumed to be changed further:

Case A) After the beginning of the negation-

Case B) Before taking a fluctuation

(positive increasing fluctuation is set by Π; see no. 202).

In both cases we obtain a varied co-moment Γ and thus a vital difference of a higher order (no. 204).

It turns from the vital series of the first order into an intermediate series or – as should be called accordingly – a vital series of higher order, the introduction of a first-order medial change and final changes (i.e. the medial and final changes of higher order) transition to the complete conclusion of the whole assertion, meaning the forming of the vital series of the first order.

218. — A further variation of co-moment Γ (see no. 205):

1) a quantitative — concerning the size,
2) a qualitative — concerning the form.

Since qualitative variation can also be reduced to a quantitative expression (no. 207), the difference for the type of analysis should not be different; but we recommend the special emphasis on qualitative variation for our special purposes.

If Γ denotes the unvaried form of the partial systematic co-moment, then the value by which Γ is to be increased is thought of in the variation denoted by $\Delta\Gamma$. In general, in case A, containing the unvaried form Γ, then the varied form $\Gamma + \Delta\Gamma$ would be assumed; while in the case B, the further division of the series will depend on whether the unvaried co-moment and the varied have equal or different times, so that as the times themselves are conceived as changeable, with respect to the initiation series, different cases of higher order can be envisioned:

219. — 1) Unvaried and varied co-moments have infinitesimal time differences due to some change conditions and result in a temporarily approximately coincidental form of work increase; the result is what we can call a complex vital series: $\Gamma + \Delta\Gamma$.

220. — 2) Unvaried and varied co-moments have different times, and thus two temporally divergent forms are set. This case yields an explicative series of vital series::: Γ, $\Gamma + \Delta\Gamma$.

221. — 3) The times differ, but are determined by some conditions such that while one change exists, the other occurs or after both are set one takes over the other; thus an explicative introduction is obtained, but at the same time one of them from which — in composite cases — a number of members again represent an intrinsic complication. This would result in a mixed vital series discharge of the most varied kind. For example, $\Gamma + \Delta\Gamma, \Gamma$.

222. — Of these conceivable cases we emphasize the one which is the most suitable for our purposes. It is from the cited second (no. 220): in other words, the case in which the initial change (higher order) strictly speaking, does not occur until the time before the partial systematic co-moment has been unchanged in its form.

The time portion then gives the preliminary section of the actual vital series of higher order. Through it, the structure of case B approaches that of A, where the form Γ is thought to be placed in front of the form $\Gamma + \Delta\Gamma$; so that at the same time Γ represents the value of the first order of the initial higher order change.

Thereby, the case should be sufficiently circumscribed to tie to our further analysis.

Fourth Section

Medial and final changes of
The independent vital series.

First Chapter

General.

I.

223. — It follows from the concept of the medial section that all changes of system C are to be reckoned with, which fit in between the introduction of the vital difference and its annulment (no. 182). Now that the realization of any change due to any current condition combination within those time limits is conceivable, it follows that any conceivable change in the system C can also be treated as a conceivable medial change.

From this it cannot be further deduced that any conceivable medial change could at the same time be thought of as the *actual mediation* for the final change of the same order; Rather, it can do so only if, according to no. 194, it can at the same time be thought of as fulfilling the formal conditions of the dissolution of the vital series.

224. — We therefore distinguish the conceivable medial changes into those which at the same time are thought to fulfill the formal conditions of vital difference annulment: the medial changes as actual intercessions of the final change; and into those which cannot be thought of as mediations in the sense indicated: as medial changes in the sense of mere intermediate changes.

II.

225. — If we approximate a system C to the ideal (no. 187), at least insofar as it can fully assert itself for a limited number of cases and within a limited period in a non-ideal environment with diminutions of its vital conservation value, thus for every complete preservation ascribed to system C, the general proposition (no. 188) for the vital series in relation to the medial section is specialized — and taking into account the results of our analysis:

If a vital series of a system C can be considered complete, then the medial section of each order must be considered continued until a change occurs that satisfies the formal conditions of cancellation of the vital difference of the same order.

III.

226. — In case I (no. 215): the assumption of unchanged conditions for the setting of co-moment Π associated with the partial-systemic co-moment Γ — in the vital series of the first order — permits the same working forms and quantities under the uniform setting of which the respective partial system has developed into a main partial system and also considered as the quantitative and qualitative increase of labor, which is now asserted. This means the unchanged co-moment simply acts as a medial change (first order); and as the system C is thought to be provided with the unchanged co-moment, and so it is — if the co-moment remains unchanged — also assured of its final change (first order).

227. — If, on the other hand, the partial-systemic co-moment does not remain unchanged but is varied, then the changes in the medial change must be thought of as changes which first raise the variation of the co-moment: this means case I goes into case II; a vital difference of higher order is set, and – assuming complete assertion – a higher-order vital series is switched on.

228. — And as in the case I the medial change simply consists of the co-moment in form, which – as a correspondingly increasing negative fluctuation – leads to the vital difference of its abolition (see no. 226), then the medial changes must now be thought of as approximating the variation of the co-moment cancellation. And as in case I, the final change of the vital series of the first order consists in just the abolition of the simple difference from the system rest point, the final change of the higher order will have to consist in the cancellation of the difference from the co-moment, i.e., the co-momentary variation.

By the final change of the higher order, case II returns to case I: instead of a higher-order vital series, the first-order vital series again occurs.

229. — Case I does not exclude that the co-moment – as the corresponding, negatively increasing fluctuation – itself can only be thought of and conditioned by any further changes. The purer that case I is supposed to be, the more these changes will approximate the value of a partial-systemic co-moment, so that their associated partial system is only able to express itself completely by serving in the complete assertion of another partial system.

230. — In general, the medial changes of the first order are essentially nothing other than as they are or can be set in the higher order vital series; so, we prefer to treat them in the higher order medial changes as best suits our purpose of analysis.   For the same reasons, we may find the kinds of co-moments generally and consider closely only the following.

Second Chapter

Selected Cases of Conceivable Associations.

I.

231. — We first try to understand our assumptions, as we have to consider the general formal conditions (no. 194) for the cancellation of a higher-order difference.

According to our chosen case, in which the setting of the partial systematic moment Π represents the case of a first-order vital difference (no. 208), the latter would find its analytic expression in the inequality.

$$\partial' == f(R) + [f(S) + \Pi] > 0.$$

On the other hand, the vital difference of higher order, which we want to denote by ∂', according to no. 204 in the inequality.

$$\partial' == [f(R) + \Gamma] + [f(S) + \Pi] > 0.$$

232. — If one thinks of the n as the main partial systems of the system C as providing a number k with a higher order vital difference, and the remainder n - k without such, then for the n main partial systems

$\partial'_1 > 0$
$\partial'_2 > 0$

.

.

.

$\partial'_k > 0$
$\partial'_{k+1} == 0$
$\partial'_{k+2} == 0$

.

.

.

$\partial'_n == 0$
Consequently for the whole system C:
$$\Sigma\partial' == [\Sigma f(R) + \Sigma\Gamma + [\Sigma f(S) + \Sigma\Pi] > 0.$$

233. — Like (no. 191) all conceivable cases in which equality is shown as :

$$\partial == f(R) + f(S) == 0$$

and inequality as:

$$\partial == f(R) + f(S) > 0$$

would apply to positive cases and vice versa, the conceivable cases inferring inequality into the equation would cover all cases of negatively increasing fluctuation. Therefore, conceivable cases of the vital difference of a higher order and their annulment could be found by an analogous treatment of the corresponding equations. But at this point an enumeration would correspond to our next task. Rather, we now must remember – following the case in no. 201 – only in a few select cases, by which the variations of inequality

$$\partial'==[f(R) + \Gamma] + f(S) + \Pi]>0$$

Considering system C:

$$\Sigma\partial' ==[\ \Sigma f(R) + \Sigma\Gamma] + [\Sigma f(S) + \Sigma\Pi] > 0.$$

Can we think of the formal conditions for the cancellation of the vital difference.

The next task would be to select and classify the changes in system C that produced the analysis from the point of view of possible *actual mediation* (no. 224); to emphasize how they can be thought of fulfilling the formal conditions of annulment; and finally it would be necessary to examine under what conditions the selected conceivable mediations could be presupposed (or not).

234. — Case A. Assume that a higher order difference is set, for example

$$\partial'_1 ==[\ f(R_1) + \Gamma_1] + [f(S_1) + \Pi_1] > 0,$$

It is thus conceivable that the environmental combination to which system C is exposed has any (at all or provisionally) direct functional relationship with the associated partial system c_1; and then the next change, which can be thought of as satisfying the formal conditions for canceling a higher order vital difference, will be one which only gives system C a change condition which is functionally related to c_1.

235. — Case B. Regarding a case B, on the other hand, in the opposite case, where the environmental combination is already functionally related to the positive incremental partial system c_1, a change that would cancel that

relationship altogether does not have to be considered fulfilling the formal conditions for vital-difference cancellation.

In case A, every change in the median will assume the value of an actually mediating value, which only offers the corresponding change conditions to the main partial system in a positive increasing fluctuation.

II.

236. — From the general case B, now that the main partial system, which is in a positive increasing fluctuation, is exposed to its change condition at all, let us set aside the following more specific cases, all of which are related to our most important case, namely, that the vital difference of higher order is due to the variation of the partial system c_1 the associated co-moment Γ_1 and the increase of the value Γ_1 by $\Delta\Gamma_1$.

It can then be restored to the equation:

$$\Sigma\ \partial' == [\Sigma f(R) + \Sigma\Gamma] + [\Sigma f(S) + \Sigma\Pi] == 0$$

1) by increasing the value in $\Gamma_1 + \Delta\Gamma_1$ in the inequality

$$\{f\ (R_1) + [\Gamma_1 + \Delta\Gamma_1]\} + [f\ (S)_1 + \Pi_1] > 0$$

by the value — $\Delta\Gamma_1$; in fact:

a) by multiplying the change condition in the opposite sense: or

b) by transferring the change itself to another partial system.

In the case of a, by restoring the equation for the partial system c_1, the equation for the whole system C will also be easily restored.

In case b, however, other conditions will be required, for example, that the partial system to which the change has been directed is either a sub-partial system or such a main partial system c_2, for which the change introduced no longer means any qualitative variation, insofar as its peculiar changes themselves already enclose it in a form, and for which consequently the increase of labor provided by the increase of growth has only the meaning of an insignificant fluctuation.

237. — The equation

$$\Sigma\partial' == [\Sigma f(R) + \Sigma\Gamma] + [\Sigma f(S) + \Sigma\Pi] == 0$$

can be established:

2) by increasing the value Π_1 by a corresponding value $\Delta\Pi_1$ — Again in this case, the above equation is restored simply by restoring the equation for the partial system c_1.

238. — 3) by the negative co-moment of the value Γ_1 in $\Gamma_1 + \Delta\, \Gamma_1$ the associated main partial system c_1 is developed back to subpartial system and instead of a former subpartial system it now becomes the main partial system, whose changes — now co-moments — varied from the earlier Γ_1 and no longer have the conditions of variation in the conditions of change. In this case, the equation for the whole system C is restored, which is the equation:

$$[f(R) + \Gamma] + [f(S) + \Pi] > 0$$

which belongs to the series of main partial systems overall and if another partial system is substituted the resulting vital value can be presented as:

$$[f(R) + \Gamma] + [f(S) + \Pi] == 0$$

Third Chapter

The conceivable changes in system C as mediations.

I.

239. — Our next task, according to no. 233 (paragraph 2), is to make the changes in the system C from our analysis which is right to make from the

point of view of the possible actual mediation (no. 224) of a negative increase in fluctuation.

Above all, we should stress the premise that such medial intermittent changes may be both dependent and independent of system C.

A change independent of system C, which can nevertheless be thought of as a measure of negative increasing variability, will be placed wherever the positive increasing variation occurred due to environmental changes and a second environmental change which in the opposite sense changes the system C independent of any set change conditions.

240. — A simple citation of this kind may suffice for a negatively increasing fluctuation of mediating changes; on the other hand, the changes dependent on system C, if they are conceivable as mediations, are to be analyzed in more detail. The types selected here can be distinguished according to a threefold aspect:

I) According to the relationship with system C itself;

II) According to its nature as a functional or formal change of system C;

III) After the time they are thought to claim their full settlement. (*vollständigen Setzung*)

II.

241. — I. Accordingly we first distinguish — in terms of their relationship with system C — such changes, which are thought of depending on system C and finally because they are completely within the same, these successful changes would also have to be experienced within system C — just as a reduction e.g., annulment of the vital difference— which, however, are thought of only in their beginning stages and not in their further course that is exclusively dependent on system C.

Such changes, — although dependent on system C in their initial terms — which take place outside of system C, we call ectosystematic; their initial members, as well as any changes whatsoever that run within system C, are considered endosystemic changes.

242. — A. Within the ectosystemic changes, we again distinguish such changes (see no. 85)

1.) The fixed dependency relationship between system C and the change condition, if they detain the direction of the change condition R_1 contained in the environment, or from R_1 discourage other change conditions that would remove, alter or destroy R_1;

2.) The dependency is permutated if R_1 changes to R_2 or changes the state of the organism to R_1;

3.) The dependency is transformed if they annul or change R_1, or if they change the dependence of system C on R_1 by accommodation of peripheral organs or by changing the distance between R_1 and the organism.

243. —B. Within the endosystematic changes, the following cases should be distinguished (see no. 87):

1) Changes due to transitory function propagation, namely:

a) Establishment of a temporary nutritional relationship by temporary variation of the physiological nutritional conditions;

b) Continuation of the introduced change from the primary partial systems to others

2) Changes due to increased or decreased exercise, namely:

a) Change in the developmental value of partial systems that are increasing in positive terms:

b) Changing the developmental directions of the system C at all.

244. — II. Ordered by their differences as functional and formal changes, the ectosystemic changes of all and the endosystemic changes of sub 1 would constitute the mode of functional change; while the other endosystemic changes, e.g., the changes sub 2, would constitute the nature of the formal changes.

245. — III. In terms of finite time, we contrast the relatively small permutations and transformations as relatively fast changes in contrast to the relatively larger, slower changes based on expressed changes.

But whether a formal change is slower than a larger permutation e.g., transformation, or vice versa, it cannot be determined without consideration of the special case.

III.

246. — The variety of conceivable changes which can form a medial (middle) section, gives rise to a multiplicity of conceivable medial sections. There are three types to be distinguished within this manifold of conceivable medial sections:

I.) pure ectosystematic;

II.) pure endosystematic;

III.) mixed

Inside each of these types exist the most diverse subtypes of medial segments depending on the specific nature of ecto- and endosystematic changes that enter the medial segment.

247. — As according to no. 219, the corresponding change over the final condition of the initial section is decided, so here is also decided the composition of the medial section. This means: if, assuming a medial section's positive increasing fluctuation, then at any time those types and subtypes of conceivable medial segments are to be assumed, which exceed any other type of medial segment imaginable in the speed of entry.

248. — It is therefore the series of medial changes than at the time of an endosystematic on an ectosystematic e.g., from a sensory to a motor, from a functional to a formal, or vice-versa: from an ectosystematic on an endosystematic e.g., from a motor to a sensory, from a formal to a functional thought, in which one change has to be thought of as occurring more quickly than the other.

249. — Is is (according to no. 225) in time, meaning while a divergence is set, an endo- or ectosystematic change is set, so the system C, under the

condition of the final completeness of the vital series, must be differentiated for so long, e.g., otherwise formal endo- or ectosystematic changes are thought to be pending until a change that meets the formal conditions of vital-significance annulment is added.

The fact that this series formation can be temporarily interrupted – by interrupting due to exhaustion – or that it can be broken by other changes of the first and possibly the second concurrent fluctuation, makes the presupposed assertion of the system C more complicated, but does not alter the general terms.

IV.

250. — Let us see now to what extent the selected changes can at the same time be thought of as fulfilling the formal conditions for the cancellation of a higher-order vital difference; still with the opinion that an unproblematic modification would also assert the discovered propositions for the first-order vital difference.

A. Ectosystemic changes.

1) There may be an existing change, contrary to the positive increasing variation, which in itself is too small to cancel it, due to the prolonged setting of its condition of summation; or as a result of an omission of a change condition, a fluctuation that was opposite to the omitted change condition can increase positively. In these cases, the fixation of the dependency ratio (*Fixation des Abhängigkeitsverhältnisses*) by keeping constant the change condition may set a change that satisfies the formal conditions of vital difference cancellation.

251. — 2) There may be an existing change which sets a vital difference or at least is not deployed and is exchanged for another, whose dependent change is opposite to the set fluctuation. Here, the permutation of the dependency relationship through the commutation of the change conditions would add a change that met formal conditions of vital difference cancellation.

252. — 3) If the vital difference is in a plus or minus set with a change condition, or in an *aliter*, so any transformation which transforms the

change condition in an opposite sense to plus or minus or *aliter*, can also bring about a change which thus fulfills the conditions for the cancellation of the vital difference.

B. Endosystemic changes.

253. —1.) a) If insignificant positive increasing fluctuations in work are set, then functional changes in nutrition with the same can cause a change in system C, which meets the formal conditions for negative increasing variability.

254. —b, α) If the current environment initially reduces or sets insignificant differences in vitality, while there exist more significant differences in vitality, only partially matching environmental changes; or

ß) Has an element of form change contained in a second change which has the original value of a uniformly set change and so had a vital difference of the second order:

Thus, by propagating the change within system C, a change can be brought about which fulfills the conditions of annulment of the vital difference, in case *a* it presents a more significant difference in vitality, at least for it's annulment and in case *ß* a form change element already contains it as a peculiarity which in the previous change contained the meaning of the vital difference.

Let's call the behavior of the system in the case of *a* the short co-moment exchange and in case *ß* give it a co-moment representation.

255. —2) The expressed endosystematic changes will produce systemic changes which fulfill the formal condition, namely, a permanent elimination of the vital-difference, in cases where the vital difference itself is based on the recurrence of a change condition deviating from the previous exercise. Thus:

256. — a) if the reccurance is a relatively uniform, the formation of the initial change to a new co-moment: a positive co-moment of the changed work increase (see no. 202); and

257. — b) in the case of a relatively uniform recurring deviation, as a result of insufficient practice, there is through negative deviations a gradual reduction of deviations to relatively insignificant differences in vitality. The

co-momentation of the previous co-moments (*Komomentierung des bisherigen Komomentes*), at the same time take over the functions of the negative co-moments and are further developed in this multi-practised set in the sense of positive increasing co-moments

258. — Such a functional takeover would be suitable: a) any existing forms of change which are relatively independent of the environment at all, which we refer to as independent; or ß change forms, which are caused by the repeating of the associated environmental constituent elements.

In the following, we denote the behavior of the system in the case of *a* as the acquisition of the co-moment, and in the case of *b* as the changes of the elements.

259. — The following case of co-moment conversion allows that the formal condition for the vital-difference annullment in the selected cases of this condition corresond to case A (no. 234);

The cases of ectosystematic mediation in general in Case B, 1, a (no. 236);

The case of co-moment representation in case B, 1, b (no. 236);

The case of the acquisition of the co-moment case B, 2, (no. 237);

The case of the co-moment change in case B, 3, (no.238);

260. —If, with the final change according to the above, a value, which is capable of canceling out the positively increasing fluctuation is first established, it can be described as substitution (in the narrower sense); on the other hand, as resitution provided that the original co-moment is specially restored.

Both restitution and substitution may be ecto-or endosystematic (i.e., conditioned in this way).

261. — Both forms and settlement, as well as medial and specified final changes of higher order are not only conceivable, but are already presupposed for system C: included in the change triggers are the forms of negatively increasing fluctuations resulting from system change forms and the setting of the secondary changes as a function of preceding positively increasing fluctuation, which are included in the premise of the triggers for

which the primary changes are to be considered as a complimentary condition. (no. 114).

262. — The trigger activations have been taken into consideration primarily as a condition for vital difference cancellation. This is not the only relation in which they can be presupposed as placed for the preservation of the organism; and there is a special relationship to be noted here, which, however, is akin to that one, but was not adequate to be treated in the same way in the context of our investigation: this is the meaning of the activation as a derivation of the above-mentioned quantities of change, so that activated movements under certain circumstances and in addition to the changes in the position of the members, continue to contribute to the conditional relationship of the environment to system C (to the preservation of the organism).

To this "motor movements" would be added and the movements of the speech organs, so a linguistic "utterance" would be expected (see note no. 6).

Fourth Chapter

The Possibility of Certain Medial Changes for Certain Cases.

I.

263. — The selected conceivable cases – meaning ecto and endosystematic changes – can be thought of as fulfilling the formal condition of vital-difference annulment of medial changes of higher order, which are relevant for us.

It cannot be presupposed that the multiplicity of ecto- and endosystematic changes just a single setting or causes one of the selected types of change or coincides with another. It also remains conceivable that ecto- and endosystematic changes occur which are either not in the positively increasing change condition or – if already – it does not change in the sense of a corresponding negatively increasing fluctuation.

Since, however, none of the selected types of change itself can be assumed to be a change condition for any partial system that exists in a positively increasing fluctuation or for the negatively increasing fluctuation of the same; on the other hand, for some types of positively increasing fluctuations, different ecto- and endosystematic changes are conceivable as annulment conditions.

Finally, such diverse changes all come to an end and could only be realized after the appearance of another change which brings the others to annulment, which is realized first, and surpasses the rest in a given time

This results in;

264. — If, in the case of a certain positive increasing fluctuation of a particular central partial system, a specific change in the medial is to be conceivable for a given case, then is must be the same,

1) a conceivable condition of change for the positive increase of as, partialsystem;

2) conceivable as a condition for the corresponding negatively increasing fluctuation;

3) conceivable as the fastest of all changes which can be assumed to fulfill the first of two conditions at the time of the vital difference.

II.

265. — As in the cited cases, a suppression of certain vital differences by a certain ecto- or endosystematic change can be considered as conceivable; In other cases the elimination of a certain vital difference by certain ecto- or endo-systematic changes may be unconceivable.

The following cases arise from the previous sentence (no. 264).

In other cases, the elimination of a certain vital difference by certain ecto- or endo-systematic changes may be unconceivable.

1) if the particular change in the case can be thought of as a positive incremental partial system;

2) while the particular change may be thought of as a condition of change for the particular partial system of positive increase, the dependent change, at the same time, does not fulfill the formal condition of nullifying the vitality, meaning it does not correspond to one of the conceivable types of vital difference reduction;

3) if the particular change may be considered as a condition of change for the case of positively increasing fluctuation of a partial system and accordingly is dependent on the change as a certain type of annulment of vital difference.

266. — Therefore, the *final* change cannot be thought mediated

1) through one of the faster forms

a) through endosystematic changes when the vital difference is thought to be greater than any of these changes;

b) with functional ectosystematic changes, when the environmental changes dependent on them are thought to be so slow that they would be overtaken by a functional or formal endosystemic change;

2) through one of the slower forms of endo- or ectosystematic changes, if one of the faster forms had already preceded it.

It is self-evident that the propositions no. 246 may appear (along with others), to be disregarded in the theoretical as well as practical demands which are customarily made of individual systemic changes.

267. — The general proposition (no. 249) can now be specified as follows: If, given a higher-order vital difference, a system C should be thought of as fully assertive, thus if no kind of ectosystematic medial changes could cause a final change, it must become some sort of endosystematic one; if no order is faster, then of a slower nature; if not functional, then to think of some kind of formal medial changes.

And, if it cannot be assumed to be of a restitutive nature, (*restitutiver Art*)the change must also be assumed to be of a substitutive (substitutiver Art) nature.

Chapter Five

The realization of certain medial cases.

I.

268. — We now search, always under the assumption that system C holds its own (within certain limits) under dimunitions of its vital conservation rule, which is assumed to be *mediating* if one of the mentioned vital differences of higher order is accepted. —

If a reduction in the vital conservation value is assumed at all, it can be assumed to be insignificant or significant, e.g., as a fluctuation of secondary or main partial systems, and a system that does not possess or dispose of substitutable forms, while it is itself still in its development or formulated state.

269. — If the reduction is more than negligible, e.g., presupposing a subpartial system or one approaching a developmental standstill without expressed substitutable forms: the anullment by ecto- or endosystematic restitution is to be assumed, on which the reduction was based, whether by instantaneous accomodation of diet or muscle tension: the first form of accomodation remains conceivable in the case in question, where a stalled systems lasting changes in its forms has become inaccessible allowing variations that should be considered significant to the extent that insignificance permits.

If considerable significant differences are assumed in addition to insignificant differences in vitality, then system C will be able to pass from insignificant to substantial: the case itself, then, is a considerable one.

II.

270. — On the other hand, the more the supposed difference in vitality corresponds to the notion of a substantial one, e.g., assuming it belonged to a main partial system of a system C still in development - the less it would be in the sense of the system's presumed approximation of the ideal, it would pass into another difference in vitality, which is assumed to have no functional relation (see no. 254). And provided that at the same time the assumed difference in vitality corresponds specifically to our selected and underlying case (no 201), e.g., especially as a varied comoment of a system already in advanced development and capable of further development - but according to its development, it is thought to contain a considerable difference in vitality, only the ecto- and endosystematically conditioned changes are considered for the formation of possible vital series, unless they are simply restituted by instantaneous accommodation of the diet or muscle tension as a final change requirement.

271. — Therefore:

If a varied co-moment in the sense of the selected case is set as a vital difference of a higher order, then the final change of its vital series can only be thought of as mediated

either:

by change, permutation, retention of individual environmental components, the entire environment, the spatial relationship to the environment;

or:

by co-momentary representation, co-momentary acquisition, co-momentary change. The first group cited includes ectosystematic types of change, the second, endosystematic.

272. — The whole first ectosystematic group, however, cannot be accepted as mediation, and therefore the final change cannot be determined by it:

1) if the environment, e.g., components, to which a spatial relationship changes, e.g., such a change, it not permitted to be maintained;

2) if you have such a change, arrested but changed, e.g., it cannot be thought of as a condition of a negative increasing fluctuation; it is

a)   that any change, e.g., adherence would only increase the positive variation; it is

b) that the development of the partial system, which is increasing in positive fluctuation, e.g., its form of fluctuation, - does not depend on any surrounding elements at all;

3) If you make a so-called change, e.g., it may also be thought of as a condition of change for the partial system of positive increasing fluctuation, but any conceivable type of endosystematic group must be thought of as being realized more quickly.

273. — It follows:

1) A vital difference of higher order whose cancellation is the result of change, e.g., conceivably holding an environmental component -, can be reversed by an ecto- as well as an endosystematic medial change; and the set of change types (no. 271) from which system C selects its medial changes is determined by the speed of the mediating change.

274. — 2) A vital difference of higher order, whose cancellation by change, e.g., holding the environmental components is not conceivable because the development of the associated partial system which is increasing by positive fluctuation -, meaning its specific form of fluctuation was not conditioned by any surrounding component at all. It cannot be reversed by an ectosystematic change, but can only be reversed by an endosystematic medial change. These cases are categorized by the type of development, meaning determined by the genesis of the fluctuations.

III.

275. — Let us turn now to the second, the endosystematic group: in the first place, the co-moment representation, as a *functional* change (no. 244) and has to be thought of as faster than any kind of formal change (no. 245), the

final change is also determined first by that co-moment point which again corresponds to the faster change.

276. — One thinks of such an initially realized co-momentary representation by the environment as once again mitigated, thus considered untenable, therefore system C (according to nos. 245, 249) will proceed to a second co-moment representation with a greater time requirement.

If however, one thinks of the other representations by the environment again and again, the system C will also change over to new and slower changes of such a representative kind until —

either

277. — a) if the varied co-moment always returns, it itself assumes the developmental value of a new co-moment and thus simultaneously fulfills the condition as an element of the change form $\Delta + \Delta \Gamma$ — including the former change $\Delta \Gamma$ — as peculiar to it, which had in the initial change the meaning of a vital difference;

278. — Or:

b) The environment has stopped varying the co-moment by form variation; the resulting insufficiency results in a fall below the value of the co-moments, that is, they are no longer co-moments; whereas forms evolved into moments for which the environmental component is no longer a condition of change whether it be that the co-moments developed independently of the environment – as independents – or because co-moments contain no other elements of change other than those which are realized provided that their condition is common to every component of the environment which is there condition of change which affects the partial system.

Case *a* corresponds to the co-moment acquisition (no. 256), case *b* the co-moment change (no. 257). The positive or negative increase in developing fluctuation values fall under the concept of positive and negative co-moments (compare no. 202).

IV.

279. — Thus in the co-moment representation, the final change is determined by the use of another co-moment already available to system C; in the co-moment acquisition through the adjustment of the developmental value of the partial system to the value of the varied moment — this includes, that Π has adapted to the new I_2 (== $I_1 + \Delta \Gamma_1$) (no. 237) —; in the co-moment change through the development of forms of conservation against variation from co-moments.

280. — Subsuming both cases of the co-moment (new)-, e.g., converting (see no. 260) under the term (endo-systematic) substitution in the broader sense:

Thus (in accordance with no. 259) in the co-moment representation a (formally) non-varied co-moment, in the co-momentary acquisition a varied partial systemic moment, in the co-moment change a combination of both values substituted for the corresponding initial set values.

We want to distinguish these substitution forms as substitutions of 1st, 2nd, or 3rd order.

281. — From what has been said and assuming that system C is complete, it follows:

If in a given case, the medial change cannot be determined as a type of the ectosystematic group (no. 272), then it is to be determined as a substitution (in a broader sense); but first as a substitution of the first order, unless it is assummed at the same time that order is so prolonged owing to the untenability of each of its members, e.g., to go beyond the development values and directions of a given system's ability to change. In this case, the medial change is to be determined as a higher order substitution; and to that since the return of the deviation had at the same time to be assumed to be relatively uniform or uneven (no. 255), a substitution of the second or third order.

V.

282. — Having abolished the variation of the co-moment, the final change of higher order is reached, closing the vital series - meaning case II returns to case I (see no. 226); the condition for the abolition of the vital difference of the first order, and thus the conclusion of the whole vital series, is obtained.

But only if there is no new variation of the co-moment, e.g., in system C the variation of a new co-moment intervenes; with the establishment of the same case I would be lost again in case II - and another new series of vital order must be completed before the system would come to rest.

In particular, it is conceivable that the removal of a vital difference of higher order of one partial system sets a higher vital series for a second as well as a first order (no. 185). Above all, the conditions for doing so would be easily realized in the co-moment representation (no. 259, case B, 1b).

283. — It will be noted additionally, as any existing form of variation, either directly or formally, can evolve into a form of conservation by means of frequent mediation, such a factor once developed, can enter into a number of co-moments and function in simple substitutions of the first order.

And from this it follows at the same time, that under certain circumstances a system can also change within persistently expressed forms (Schutzformen), e.g., for example as is one depending on environmental constituents independently substituted; or vice versa.

A transitional form our general assumption includes the case of mixed forms of expression, where - for the time being - both genus of expression still exists together.

Fifth Section.

The Final State of System C as Member of Independent Vital Series.

First Chapter.

The Approach of the Final State.

I.

284. — After we have sought to determine the vital series of higher orders in their composition and course, we turn to the determination of the final qualities which enter into the members of the vital series.

If the type of vital difference allows several conceivable kinds of medial changes, then (according to no. 273) the shorter time that is needed to settle determines the outcome. If, however, each member of a vital series is taken as a majority of conceivable system states and the entire series is run in time, then if any arbitrary compliment condition Kx is given at any one time, those system properties must be thought of as successively composing the series, and the setting of the series requires a smaller amount of time each time.

285. — The successive final state would be determined by a smaller time for each system in which the transition from initial condition to a plurality of conceivably mutually distinguishable end states; but this general concept can be determined by a kind of difference of the "system C" from other "systems" whose general characteristics *of conceivable transitions* still apply.

Namely, it is conceivable from any initial condition to transform to several final states, thus for a constant Kx the transition to the end condition as the lesser timepoint and can be thought of as correspondingly most closely approximated to the initial state of the moment in question.

286. — This means:

System C can be thought of as the setting of a complimentary condition Kx from any initial condition in each timepoint to the final state, which at the relevant time must be assumed to be the one closest to that of all conceivable end states.

287. — Consequently (in connection with no.249): Should a vital series of a system C be thought of as complete, so must the changes in the median be thought of as long as possible from the nearest (at the time of the addition of the complimentary condition) to the more distant final qualities, until an end condition (*Endbeschaffenheit*) is followed, which fulfills the formal conditions of the cancellation of the vital differences.

II.

288. — Let us now look for a suitable and sufficient selection and classification of types for our purposes, in which a system C of a certain functional end condition can be approximated and so we start with the simplest case.

1) As such, the case appears that an end condition is set at the time τ_{n-1} which was previously set in the immediately preceeding time τ_{n-1}. In this case, the system C state was absolutely approximate to the time τ_n, and its value parameters of change with respect to this are set to 0 at time point τ_{n-1}.

289. — 2) On the other hand, take the case that the systematic final condition at timepoint τ_n is a repetition of a final condition (end point) which was set in a past time τ_{n-1} and in the meantime has already decreased, so the change of the system at the timepoint τ_{n-1} with respect to the final condition set at timepoint τ_n will be all the less, the less the system τ_{n-1} has been removed from its original condition by its decrease.

It was, for example, the decrease in the system end use work due to opposite changes called fatigue, exhaustion, dormition (*Ermüdung, Erschöpfung,*

Einschlafen), so the less fatigue, exhaustion, dormition, the smaller the change becomes.

290. — 3) If more generally we think of the decrease — a much used expression — *as a function of time*, then the shorter the period of elapsed time after the initial condition is set, the smaller the change will be.

291. — 4) If you think of finality as only a partially temporary change (no. 110), it is the limit of its decrease in remanence (no. 111); the greater the remanence, the greater the restorative approximation and the small the change in repetition.

292. — Or more generally:

— 5) The approach will be greater in such cases where the smaller the changes, the larger the set remanences or in other words: the larger the sum of associated remanences.

293. — 6) And finally, system C can be approximated to a formal and functional end state, that is due to a system C line of development which coincides either with a typical general developmental form of system C (e.g., puberty) or with such forms of growth and training in which a specific system or expression of system C applies; or also with forms of pathological changes of the system (nos. 103, 106).

294. — The cases 1 — 5 have in common the same finality already set in their form. On the other hand, it is peculiar to the cases mentioned in subsection 6 that their final condition had not yet been set and — permit the expression — was thought to be an *eruption*. We also want to distinguish these two types from each other by designating the former cases as repetitive and those of the latter as primitive approximations (an initial condition for an end state).

295. — It follows that every repetitive approach must be thought of as originally primitive, since every repeated change, in order to be repeated, had to be set for a first time.

III.

296. — All of these selected modes of approach are conceivable on their own and in relation to system C insofar as these approaches fall under *prepatory* changes (no. 102), to be included in our general requirement: the terms most approximate and prepared are substitutable.

297. — The substitution gives the following:

If several final properties of system C are conceivable when setting any complementary condition Kx as members of a vital series, these final qualities must be thought of as composing the same, which are thought of as the most prepared with respect to Kx and the time of each settlement.

298. — What was said in no. 287 about the progression from the closest end state, also applies to the progress of the most prepared to the least prepared changes and ending states.

299. — It follows:

If a system C is thought to be asserting itself under the diminutions of its vital conservation value, it can be thought of as asserting itself only within the scope of its preparation.

IV.

300. — If (according to no. 297) the setting of an end state depends on the relatively largest provisioning, then the setting of the end state must be contingent on the modes of preparation attributed to the system.

If only one type of preparation is permitted, then the most prepared of our final cases depend on the provisioning of the lowest acceptance of an already set end state e.g., depending on the fastest repetition; or the greatest remanence, and in this case again the frequency or the sustainability of the previous changes; or the progression of development in the case of a predisposition or exercise, in the second of a typical instance, in the third case of a pathological processes.

301. — If, on the other hand one thinks of several or all types of states in a system C as concurrent, then the setting of the final condition can not depend on any of the individual components, but must be dependent on the totality of the same, meaning – the results.

If at the same time, one thinks of a multiplicity of complimentary conditions of different kinds, then the system change or the end state of each time point is the result of a higher order, e.g., results derived from several concluding states.

V.

302. — As an end state can be conceivably approximated by setting prepatory changes of an initial condition, so it can be considered that changes can be made to the initial condition which contradict the modes of preparation; e.g., by degeneration, cessation of the psychological or pathological, typical or non-typical development, delay of repetition time and by reducing the expression in any way.

Second Chapter

The Constitution of the Final State.

I.

303. — We now turn to the general analysis of the final form properties of system C, which are placed in the vital series, and we must pay attention to their relationship to the environment as a condition of their settlement as well as their ability to be set (*Setzbarkeit*). Due to the large changes in the initial environment, system C, in our general condition, is exposed to many varied change conditions; but these are not thought to occur all at once as complimentary conditions (no. 105).

304. — Let's assume for the sake of simplicity, that for all environmental components as conceivable complimentary components, that system C as the epitome of the systematic preconditions (inherited preparation value, sustainability of system changes etc.) is the same, so from the competing conceivable complimentary conditions, the most used of them becomes expressed. We can accept the assumption above — if you will — as fiction — made easier by not considering the epistemology of individual differences of accidental inheritance, sustainability, special circumstances etc.

We can take the above assumption, if we will, as fiction, because, for a general epistemology, the individual differences of accidental inheritance, sustainability, special equipment, etc. are not necessarily considered.

305. — Now the conceived competing complimentary conditions become the ones that are most likely to become expressed, which are more practiced and so the environmental components most likely to become true complimentary conditions are those which are mostly prevalent in the environment.

306. — If one thinks of these more frequently recurring environmental constituents with variable composition — e.g., as variable combinations of the environment (in accordance with our empiriocritical assumption no. 21) — then the initial expressed complimentary conditions must again be composed of those elements comprised of the more frequently recurring environmental combinations which themselves have most frequently reappeared inside these combinations.

307. — Therefore:

If, in the competition among the environmental components, system C is considered the epitome of the systematic prerequisites and is consistently assumed to be the equivalent, the initially expressed conditions are composed of the constituents most frequently recurring within the environmental combinations which are set repeatedly.

308. — This sentence is applicable to cases of the setting of a repeated single environment combination under variations, as well as the case in which several environmental combinations are provided with the same successively set elements.

309. — Let's denote the final endings which are thought to be conditioned by the environment as generally dependent (as opposed to the independents: no. 258); those caused by an environmental combination especially as dependents of the lower order as well as those containing several combinations of environments which themselves are dependents of higher order; — Let's conclude:

The form of the system C changes and the associated final end state are thought to be exclusively dependent on activity, meaning that the form of the final state caused by complimentary conditions is also conditioned by the most frequent activity, e.g., recurring, be it an individual constituent, be it a majority of combinations of arbitrary environmental constituents, — it must be considered conditionally — the form of the original dependents of any order must be considered conditionally as the repetition of a multiple set of environmental combinations.

II.

310. — Now, as the finite state forms were confronted with so many complementary conditions in their formation, but also repeated elements or which were, in shorter terms, related, it follows:

Every initial end state is thought to be a relatively repeatable one.

311. — It is also thought to be the case with the setting of the final form state, that the environment is to be considered repetitive as a multitude (*in dem Masse*) — as a recurrence of the relationships is presupposed — and thus remaining an environmental combination or a combination majority containing a sufficient number of repeating elements. It follows from the multiplicity of settings that the repetition of an end condition form in relation to the various environmental combinations is also conceivable.

312. — The majority of environmental combinations in which the setting of the end states can be conceived as repeatable, will be referred to as the associated combination sphere. — In contrast, an end-property form (*Setzung einer Endbeschaffenheitsform*) is referred to as a multiponible; and we distinguish dependents of different orders as we do multiponibles of different order.

313. — If the multiple settability cannot be thought of as dependent on the environment alone, but depends also on the entire conditions within the system C itself, then the end-property forms which are in some sense independent of the environment — also independents (no.258) — - can be thought of as multiponibles.

Third Chapter

The Change of the Final Forms.

I.

314. — The determination of the dependents as multiple-settable end property forms (multiponibles), however, allows the possibility of a change in the same. Our general requirement includes three cases of this kind:

1) By the variable position, which the system C can take to the surrounding combinations, or these with one another, the case may arise — without change within the specific environment combination itself — that a change occurs only from its environment, even only a transition to a particular combination of complimentary conditional surroundings in a direction other than the most frequent.

2) After the dependencies have developed such as lower or higher order, a change within the associated environmental combination is acquired or concerning the environmental combination spheres and the meaning of a complimentary condition.

3) An unset environmental condition which becomes a complimentary condition.

315. — In cases 1 and 2 an earlier complimentary condition is already set, and whatever somehow still receives the meaning of a complimentary condition, can only — if it is assumed to belong at all — be thought of as an addition and to which the originally set complimentary condition is multiplied.

316. Since the end state form was set by the original complimentary condition, it is reset in cases 1 and 2, and the complimentary condition must be thought of as necessitating a change of its original final corresponding form.

317. That means:

If, as a result, of a changed complimentary condition, a different form of end condition than was set formerly is required, then such is always considered only as a change of the earlier end condition forms.

This theorem also applies to a special case that a certain part of the single environment combination, which is not repeated in the related environmental combinations and developed a partial form of the original end state; that is a partial form was set that was not conditioned by the repetition of a majority of environmental combinations. — The above sentence also applies to the special case, that development is not thought to come from the environment, but from system C itself as the epitome of the systematic preconditions.

318. — For the remaining case 3 (see no. 303):

If an environment combination is allowed as a complementary condition at all, and as such has not yet been set, then it can first be thought of as a real complementarity condition, as far as it is related to any previous complementary conditions; meaning it coincides with a conceivable complimentary condition, which was already a real one.

319. — The realization of a complimentary condition that has not been established can be understood as a decomposition of the complimentary condition into two components: in an already prepared and an eventual addition of the same.

Herewith, case 3 returns to the first two cases, and the proposition asserted (no. 317) for those cases also applies to case 3.

320. — It follows:

If the end conditions are considered being dependent on the activity, no end state is to be thought of as an absolutely new form, but as either a mere repetition or as a variation of repetition.

And vice versa, any innovation within the final state would only be thought of as a variation of existing ones.

II.

321. — If, therefore, a changed system condition form is set, at the same time the unchanged is always presupposed.

If one thinks of the setting of a modified end condition form only the expressed or exercised gain comes into consideration, so you have to think of the unmodified as a part existing before the changed part.

322. — Consider the advantages of minimized activity, e.g., the complimentary condition is greater than the original state, the time difference minimizes to the null and the changed final condition can be set immediately.

Both cases do not exclude a changed final state returning to the former unchanged condition (no. 221).

323. — For the special case that the unchanged end condition it is synonymous with the unvaried form of a partial system co-moment, in the case of no. 321, we would obtain the composition of first members of the vital series (no. 220):

$\Gamma_1, \Gamma_1 + \Delta \Gamma_1$

324. — Consequently, provided that $\Gamma_1 + \Delta\Gamma_1$ is at the same time the initial change of a higher order vital series (see no. 222), this latter is itself (no. 320) in its preliminary section as pure repetition of a partial-systematic co-moment (Γ_1), and in the initial section as a variation of the repetition (= setting of the varied co-moment $\Gamma_1 + \Delta\Gamma_1$) and — after manifold change of pure repetitions and variations in the medial section – finally, it is to be determined as the nullification of the variation in the final section.

III.

325. — Let us denote an end property form, which and insofar as it cannot be set several times for an individual, is at the same time unchanged as a subconstant (*Subkonstante*); it follows finally:

If the higher-order vital series begins with the variation of a partial-systematic co-moment and ends with the annulment of the coincidence variation as such, the multiponible approaches an end-texture which is no longer varied by the associated environment combinations for the system C in question , that is, it can be set unchanged (within certain limits) with respect to the latter and to the individual system C, which means that the multiponible approaches that of a subconstant.

Fourth Chapter

Review of the determination of the vital series and fluctuations.

I.

326. — To conclude our analysis of the higher-order vital series, let us take another look at its individual provisions and their relation to the general determinants of fluctuations.

The higher-order vital series we select is composed of the principle values of our no. 201 case. First:

From Γ_1. Hereby is designated an end condition of the system C - first in shape; the final state form coinciding with a uniform increase in work which acquires the significance of a partial-systematic co-moment (no. 202) as the special end condition of a partial-systematic co-moment conditioned by the environment again dependent on the dependencies (no. 309), on its composibility after a multiponible (no. 312) and as a partial-systematic co-moment, an unvaried practiced fluctuaton. Furthermore, from the change $\Delta\Gamma_1$ by which Γ_1 is increased positively or negatively; where (according to n. 206) $\Delta\Gamma_1$ can also be due to the set end condition of another comoment. We then assume (in accordance with nos. 321 and 323) that the associated change times are distributed by the different exercise in such a way that Γ_1 is realized before $\Gamma_1 + \Delta\Gamma_1$; and we accept further (according to no. 224) that that a number of switched-on (untenable) and even possibly again varied endings alternated with the varied final condition before the final change (higher order) $\Gamma\omega$ had been won, which, depending on a restitution or a substitution in the broader sense (according to no. 280, see no. 260), as $\Gamma\rho$ or $\Gamma\sigma$.

Incidentally, Γ_1 will still be able to co-appear after its retirement in the preliminary section — totally apart from the Medial section; but for simplicity sake let's refrain from this and other complications, which do not cause fundamental differences. Likewise, we don't take into account the finite forms independently developed by the environment, because the difference is not yet considered here.

II.

327. — The relation between the given individual definition of our selected series of higher vital orders and the general definition of the fluctuation (see no. 162) is illustrated by the following table, in which we consider the vital series of the first order as sufficient to indicate the relation to that of a higher order and on the other hand represent the greater variety of values in the medial section by Γ_1, Non- Γ_1 for the sake of simplicity: even the series Γ_1, Γ_2, Γ_3 ... Γn would not be sufficient to express the approximate medial changes since the values Γ_1 and $\Gamma_1 + \Delta\Gamma_1$ can interact in a way that cannot be determined from the outset.

Compare the graphical representation of the fluctuation of system C in the attachment.
(see Note no. 7)

Case 1: Vital Order	Initial change	Middle change				Final change
Case 2: Higher Vital Order		Sub-section	Initial change	Middle change	Final change	
Overall vital difference	1st Order	1st Order	Higher order	Change	1st Order	0 Order
Members of the series	Partial system movement Π	Γ1	Γ1+ Δr1	Γ1 Non- Γ1	Γw= Γe od Γw Γo	Annulment of fluctuation
Fluctuation size	The system O only enters into the conditions of being awake after the fluctuation has been set	Sufficient for significance	Sufficient for significance	Sufficient for significance	Sufficient for significance	_____
Fluctuation form	...	As a multiple, it depends on the repetitive components of the environment	Reduction of the multiplicity with respect to the previous associated environment combination	Change	Approach to a sub-constant	_____
Fluctuation relevance	...	The substantial fluctuation belongs to a main partial system	The substantial fluctuation belongs to a main partial system	The substantial fluctuation belongs to a main partial system	The substantial fluctuation belongs to a main partial system	_____
Fluctuation direction	...	Fully negative	Changes in positive direction	Change	Cancellation of directional change	_____
Fluctuation variation after _____ - form	...	Partial system fully trained component	Positive trans-exertion	Change	Negative trans-exertion	_____

Fluctuation variation after _____		Maximum proficiency	Decreased proficiency	Change	Increased proficiency	
- value	...					_____
Fluctuation variation after _____		Minimum articulation e.g. opposition	Maximum articulation e.g., opposition	Change	Decreasing of articulation e.g., opposition	
- context	...					_____

Sixth Section

The Higher Order of System C.

First Chapter.

The Higher Order Setting of System C.

I.

328. — We still have to purse the little path of analysis (no. 127) we opened a bit further: after we have analyzed the changes of system C according to the significance of their assertion with diminutions of their vital conservative value in the sense and scope of our purpose, their significance should also be considered for the assertion of the systems of higher order – meaning also for considering system C of other individuals.

Every system C viewed from its partial system is a system of higher order because it is composed of partial subsystems; and every partial system becomes a higher-order system when considered as a compound of composite parts; and these composites are themselves formed of composite parts. We have to take this consideration for our view of system C.

329. — Assuming a system C was fully asserted under the diminutions of its vital conservation values, it would be labelled as follows: assume that it is a system of systems (*System von Systemen*) — and this is so if one of its partial systems has experienced a diminution of its special vital conservation value.

Of the various forms of assertion of an individual system C, the case now considered is one where the change is from the positive partial system c_1, to other c_2, c_3, ... and finally, with a change of one of the secondary systems, the formal conditions of vital-difference annulment are filled for c_1.

330. — These changes which are added can again have an opposite meaning in relation to the vital conservation value of their associated partial system:

they can either help the partial system to a new vital difference e.g., multiplied, or as such be repealed e.g., reduced.

With respect to the former, a special case of assertion occurs that will ultimately wipe out — including the partial system itself — the whole system. In the latter case, the special case of partial system assertion expresses itself by asserting one or more of the others. In the first case we have a form of the partial assertion of the system C, which threatens the preservation of the whole system; in the second case, a mutual form of system assertion, which we have already encountered (no. 229).

II.

331. — Let us proceed from considering of a single system C to the assumption that two individuals exist as M and T, and two systems exist CM and CT that are functionally connected, each with a vital difference ∂M and ∂T, such that M is connected to T and T to M generally, but also specific medial changes following the setting of ∂M are related to T in terms of change conditions.

Thus, M and T (after no. 41) together form a *system* in which a multiplicity of changes in the individual can be thought as being conditioned by the system CM.

III.

332. — Then the meaning that T has as a change condition again for M can be reduced e.g., is picked up, by modifying or preventing M from changes, e.g., messages of T are suppressed, or removing system changes from CT on which the movement and communication was based or by removing T from the environment of M (trapped or expelled in a certain space or even destroyed); or conversely, the meaning which T has as a condition of change for M can be fixed and increased: it can be approached by the

ectosystematic changes of system CM as M is held in within its environment, it may be the movements and notifications, e.g., the conditional changes of the system CT being conserved and increased — T itself then may be obtained by medial changes of M.

III.

333. — Each of these conceivable changes of T caused by M presupposes another form of ectosystemic change of the system CM as its condition: of all conceivable ectosystemic changes of system CM, the most prepared ones would be set, that in further development would acquire prepatory advantage, and whose setting conditioned the change of T which then cancelled the vital difference ∂M. Thus, of all the ectosystematic medial changes of the system CM, which are at the same time conditions of change for T, in the further development of the system CM those would be obtained which would be able to change T in the sense of the formal conditions of the cancellation of the vital difference ∂M.

334. — Of all conceivable ratios of canceling a vitality reference ∂M by a change of T, however, in the long run only one will be able to be obtained in which T is also preserved.

335. — T itself will finally find, within such a relation, the most favorable conditions of its preservation if, on the one hand, this relation not only satisfies the cancellation condition of ∂M, but the others: by canceling ∂M at the same time a vital difference of its own in system CT thus cancelling ∂T; and on the other hand, if a vital difference ∂T also causes a vital difference ∂M, then (∂M) itself can be cancelled by ∂T. This means: if the assertions of CM are so functionally linked to the condition of the assertion of CT, that the increase of the vital conservation value of CM is due to the multiplication of CT, and a vital conservation value of CT at the same time causes the same for CM.

336. — The less the system CM relies on its vital differences by setting such in CT, the less the system CT grows its own vital differences through CM, but from the rest of the environment, which may be called non- CM; and the more the vital differences set to the system CT also set vital differences in the system CM, which would be canceled by the elevation of those set at CT, then the more the vital differences arising from the system CT from non- CM become communal. Joint medial changes of which those changes which vary non- CM or the relation to non- CM in a correspondingly opposite sense (see no. 252) would be able to cancel the common vital difference.

337. — Now, whatever applies from M to T also applies from T to M and as for CM and CT, as well as for more than two systems C, but such systems whose alterations depend on each other are themselves (according to no. 41) under the notion of a system, and hence from the single system C, and are to be described as a system H of higher order or ΣC:

A ratio of the difference in vitality between two or more systems C, and hence a system C of higher order, will find the more favorable the conditions of its preservation, the more the difference in vitality is a mutual one. The higher-order vital differences only increase from non-ΣC and thus have acquired the importance of (a kind of) joint liability.

338. — On the other hand, the conditions for the preservation of a higher-order system C will be all the less favorable: the less the relation of the assertion between the individual systems is a reciprocal but a one-sided one; meaning that the less the vital differences of the individuals come from non-ΣC, but derive straight from ΣC; The less the vital differences arising from non-ΣC are communal but are individual and thus isolated (non-communal).

IV.

339. —Such systems of higher order, whose parts are systems C, are to be presupposed everywhere, and where the movements e.g., messages from individuals, by which they assert, set or abolish vital differences of another

individual, so that also the medial changes of the second associated systems C are change conditions for the first associated system C; therefore meaning turning to the condition of the setting and removing vital differences; as in every smaller or larger human society. Such systems of higher order comprised of human individuals e.g., the systems C of humans, may — in the absence of another, sufficiently relation-free expression — be called congregational systems (*Kongregalsysteme*)(of a different order).

340. — And in accordance with what has been said (no. 337), we call every individual assertion that, taken as a condition of change for the other individual systems enclosed by the congregational system, as favorable to the preservation of the congregational system, as positive or in other cases negative; and higher order systems C or assemblages with predominantly positive individuals are considered positive and in the opposite case, negative congregational systems. The relation of the condition of change itself, in which the individual systems stand with each other, if it is considered from the point of view of the congregational system, is allowed to be referred to for a short time as congregationality (*Kongregalität*).

Second Chapter

Conservation of the Positive Congregational Systems

I.

341. — From what has been said it follows, that — as far as the congregationality (and not the power of particular external events or the special ability of the individuals or of the race to abolish the vital differences at all) is concerned —

It follows then:

As the congregational systems, which are maintained in the course of further development, only the positive congregational system is assumed; whether the negative congregational systems were destroyed by inner self-dissolution or by external events it could be that the negative systems transformed into positive ones.

342. — The conditions for this transformation are provided in the following conditions:

1) The changes caused by the individual system CT, which initially have the meaning of a cancellation of vitality difference for a second individual system CM, acquire more and more meaning the more they have a vital difference set in the typical development of the CM system: to develop in CM a functionally and formally determined main partial system (no. 115), which would be in a dietary or labor fluctuation (no. 192) then (referred in no. 203) in a first-order vital difference if CT were not set. But the more this is the case, the more all changes, which means a reduction in the vital maintenance value of CT, provided that they are set the same — it will not be misleading to call — *congregational* moments or variation moments of CM, also for this individual system a vital difference of the second order, to whose modes of suspension belongs again the cancellation of that reduction of the vital conservation value of CT and thus the increase of the vital conservation value of CT. It forms with this a possible earlier one-sided relationship of the vital difference suspension into a definite mutual one (relationship).

343. — 2) Furthermore, the more a system of higher order, in the sense of positive assemblage, develops in the manner indicated, and the fact that the changes of the individual systems acquire for each other the importance of abolishing a first-order vital difference attached to them as a positive congregational systems, the more changes will be made to the assemblage comprised of a single system, which differs from those on which the *congregational* partial systematic moment or co-moment of the other is based. It also differs for the vital differences of the second order for the others.

II.

344. — For the abolition of these deviations, which cause differences in the vital differences of the second order, the following conditions are assumed:

1) In the event that the deviating changes belong to a still sufficiently varied individual system:

a) the changes of the other systems act as exercises, and the individual system practiced by the other systems varies the totality of its systematic change preconditions by adapting to the practiced changes (adaptation of the matches); or:

b) the rest of the systems act as regulators, and the deviations from the individual system are varied by the other systems in the sense of agreement, reducing or suppressing the totality of the systematic change prerequisites for the deviating changes (abating the deviancy; no. 332).

345. — 2) In the case that the deviating changes belong to an individual system which can no longer be sufficiently varied: the vital differences of the second order conditioned in this way are not canceled out by varying the entirety of the alteration preconditions of the individual system, but rather by varying the dependency ratio in which the other systems stand in relation to the deviating one, e.g., is canceled — in the spatial isolation or distance, e.g., the ectosystemic conditions are presupposed in the annulment of the system in question.

346. — 3) For the case that if the deviation is based on a further development of the positive assemblage system itself, which occurs earlier in one single system than in the others, the deviation can also be removed by the variation of the other systems which gradually acquire the specific deviant form of change as its own.

347. — In all these cases the vital differences arising from the individual systems of the positive assemblage system ΣC are not only abolished, but the

condition of their setting is eliminated altogether, and thus the condition for vital differentiation of the single system belonging to a positive assemblage system ΣC is limited to non-ΣC.

348. — By limiting the conditions to vital differences for the systems enclosed by ΣC to those set by non-ΣC, the vital differences caused by non-ΣC finally become more common, and more and more individual and the more non-ΣC itself has the meaning of a community environment (*gemeinschaftlichen Umgebung*)(see no. 336).

III.

349. — As for conservation, the conditions for the growth of conserving congregational (assemblage) systems are presupposed in our analysis; here only two may be highlighted:

1) growth through the excess of births over deaths and immigration through emigration;

2) growth through the involvement of another congregational system or through the combination of several into one higher-level congregational system. In both cases, two congregational systems that behave neutrally or negatively, be it initially unilateral or mutual, and transform negative into positive contexts.

IV.

350. - How the vital difference of the partial systems of a system CM is abolished by the removal of the vital difference of corresponding partial systems of another system CT, and how this preserves the conditional congregational system (but also of the more comprehensive system to which the latter may belong) is shown by sexual intercourse, when analyzed purely on its physiological side with respect to system C. From the same point of view, then, the formation of marriage further shows a development of that

traffic; of the congregational system set by it, to ever higher (and more lasting) positive congregational values. Outperformed in terms of generality and positivity, the development of congregationalism is still in the (related) case that CM belongs to a mother, and CT to her child. In a majority of children, their systems C can re-enter into the relationship of mutual vital difference cancellation and with other positive congregational values due to close coexistence, which then, in conjunction with the higher development of sexual intercourse bound to marriage, the family becomes the firm foundation of a positive congregation (assemblage) for the development and maintenance of the tribe.

V.

351. — In retrospect, for the partial systems, if they are subsystems of system C e.g., for system C and CM, CT and so on, and if they are subsystems of congregational system ΣC, e.g., for ΣC itself, we should in general note:

The more a subsystem asserts itself by reducing the vital conservation value of others belonging to the same overall system, the less favorable are the conditions for the maintenance of the entire system: The more the subsystems assert themselves in the sense of mutual increase of the vital conservation value, the more favorable are the conditions for the preservation of the entire system.

Conceivably the most favorable condition for the preservation of the whole system would be if no subsystem asserted itself by diminution, but each by increasing the vital conservation value from which it asserted itself; so the *perfect ratio* would be designated as follows: each subsystem would fully assert itself under the greatest possible increase in the vital conservation value of the greatest possible number of other subsystems, and thus also asserted itself in the total system itself under the greatest possible increase in the vital conservation value of each subsystem.

Seventh Section.

The Variation of the Independent Vital series through the Further Development of the System C.

First Chapter.

The Variation of the Vital Series.

I.

352. — Since, like every other form of change, so also is a whole series of changes, which is set in any time, and may already have been set or not set at an earlier time, the setting of a vital series may also be considered as a first or repeat and in the latter case may be thought of as an unchanged or altered one. For as the presupposition and certainty of a series of vital functions presupposes a development of the system C, a variation of the vital series through the further development of the system C is to be presupposed. This expands the task of our analysis to all conceivable variations, insofar as these are conditioned only by the further development of the system C; Thus, the narrower purpose of our investigation allows us to confine ourselves to some more general and important cases, concerning, on the one hand, the variation of the "no" as such and, on the other hand, the variation especially of the final change: in both cases the series and final changes of higher order are meant.

Since in the proceedings an absolute disregard of the further development was not well-advised, individual cases of the variation of the vital series by further development of the system C were already pointed out; for example, in the transition from indefensible to durable final changes (see no. 325) and in the interchange of forms of protection dependent on environmental components and one of them independent (see no. 283); other cases were already included in the premise of various preparations of system C, as well e.g., a series of vital signs, each of which is thought to be the result of any number of competing forms of preparation (see no. 301), may already be thought of as being composed of different endings, merely as a result of the variations of the competing preparations, assuming the same complementary conditions at different times.

II.

353. - We begin with the variation of the vital series as such.

1) An end condition which produced a final change may at the same time take on the value of a very lasting change (*sehr nachhaltigen Änderung*), so that the time of change necessary for its setting is reduced to such an extent that it changes the other end qualities formerly between it and the initial change, in smallness it surpasses and thus is closer to an earlier repetition, possibly directly up to the initial change time. In this way, the series is also concluded for the assumed case, before further setting previously switched-on medial changes: these are therefore excluded in the present case from the vital series. The latter itself has been abbreviated accordingly.

III.

354. — 2) If one looks at specific sustainabilities as well as peculiarities of a typical individual or pathology e.g., from their development, and if one only accepts exercise as presupposed by the more or less frequent setting of environmental constituents as complementary conditions as a condition of the vital sequence, (see no. 304) for the initial setting of a series of vital signs, it follows that the initial change set after the setting of any complementary condition is also pure as it is the change of the system most frequent with respect to the complementary condition, and conversely, that with respect to the complementary condition the most practiced change is to think of the initial change of the system.

Under the same condition (of the mere exercise), the composition of the vital series must be conditioned by the transition from more to fewer changes (according to n. 298); but always the last end condition must be thought of as the initial condition to the next.

355. — If, the setting of a series is thought of as a repetition of it, each member of it, the more frequently the series is recollected, the more practiced, and consequently the preparatory value of its members is all the

greater, and consequently the period of change of time required for the setting of each system end condition becomes smaller.

356. — Consequently the individual endings of a series are set one after the other more rapidly, the more the members of the series are crowded together, and the faster the series runs, and the more often it is expired.

357. — If, one thinks of times becoming even shorter, then the pushing together of the individual members can be thought of as merging into one another, as long as the previously set members remain set while the later ones are set.

358. — Equally, however, an end condition can be thought to have been skipped if, as a result of the shortening of the time, the series of intermediate links continues too quickly to a more lasting final condition — the time for the pertinent intermediate links becomes infinitesimal.

359. — Thus, through exercise, the vital series becomes thickset.

IV.

360. — 3) Even in the cases (see no. 301), in which the initial setting of a series of changes is not exclusively conditioned by exercise, but every change is thought to be the result of several competing types of preparation, the repetition of the series may be considered in the case of practice, as approximated to the extent that equality and differences in repetition — as exercise multiplication or reduction — cause changes in the vital series.

361. — If one thinks of a vital series, the first settlement of which was based member by member on the competition of several types of preparation, but, after it had been set, it was repeated in the same way, then the same variations must be conceivable conditionally, in the case of simple repetition.

362. — However, if one thinks of the repetitions (after remarks no. 352), as different endings arise, then again one can be thought of more often than the others.

If the final finishes are thought to be more frequent than the others, the finals will be repeated more or less frequently after their initial setting, e.g., and thus subsequently thought of as more or less practiced.

363. — As a result of the subsequent variation of their original preparatory values, which were done with this subsequent major or minor exercise, variations may again be considered which are related to those we have mentioned in no. 353:

Final end states originally located away from the initial change may approximate it; others, who were originally close to the initial change, move away from it. Thus, two final end conditions can also interchange their positions with each other.

364. — In this way endings, with which vital-difference suppression is set, can again be classified before those with which they were not or not yet set: and thus also endings which previously belonged to the vital-series, are displaced from it and thus eliminated.

365. — But also an end condition, which originally did not belong to a vital series, can be contravened and activated by other activities or by a different preparation into the vital series.

366. — If the difference of the associated change to the end state change times on which the activation, e.g., if the elimination of end-properties is based on the same idea, — then the activation and deactivation must also be repeated when the complementary condition is repeated.

367. — The more frequently an end condition of a vital series has been excluded as a result of its lessened exercise, the more rarely has it been expressed within the series, and consequently its preparatory value continues to decrease and its change time increase further; and the more often an end condition has been restored as a result of its more frequent exercise, the more frequently it has been practiced again, the more its preparatory value continues to increase and its change time continues to decrease.

368. — As by the reduction of exercise thus an end condition can be thought of permanently switched out of a vital series, and the final condition is no longer considered as an appendage.

This case of extreme exercise withdrawal will occur when system C, after completion of the vital series, passes over to end conditions which are more experienced than members of other vital series that are switched off.

369. — But it is also very conceivable that an end condition originally belonged to several vital series or is inserted later; In these cases, an end condition with respect to a vital series may be an appendage with respect to a second integral part — may be either permanently suppressed as regards a complimentary condition, or be constantly maintained with respect to the second part.

370. — But if an end condition is not classified in any other vital series or has been eliminated (see above, no. 368), then its preparatory value may be reduced to a minimum by lack of exercise — its formal determination can be reduced to a great extent (*wither away*).

371. — If the elimination of endings is based on the competition of differences in exercise or of various forms of preparation, in any case the exclusion from a vital series can only end up with no endings which are superfluous to abolish the difference in vitality-be it because they are such an abolition altogether were not suitable, be it because others had forestalled them.

372. — Therefore:

a) If, from a complete series of vital end conditions states that originally belonged to them are later eliminated, then those who have been turned off can only be thought of as endings, with the setting of which there was no (not yet, no more) vital difference, which was unnecessary in this sense.

b) As a result of the elimination of expendable members, a vital series can be thought of as being more and more restrictive to such endings, which are not expendable for the purpose of vital-difference-elimination, and are therefore indispensable in this sense; namely, any limitation is to be thought of in the direction of those endings with which the vital difference was most quickly set aside.

c) As a result, the vital series has become more successful in its medial section.

The prerequisite for these shortenings, simplifications, and restrictions is always that the systems are still sufficiently viable and not already frozen.

V.

373. — One finally thinks of any arbitrarily set of medial changes freed from all expendable members — that is all members that did not exist in a more ideal sense than in actual mediacy (see no. 224) — and limited by their elimination, as well as by the compression to a minimum of their duration;

Such a medial series, which would have reached the extreme limit of varying the number and arrangement of its members, as well as its expiration, would be considered perfect in view of bringing about the abolition of the vital difference in the shortest time period with the least medial changes which may be called *pure mediation (vollkommener Vermittelungen)*.

374. — If the conditions of the expressed series variation — and their different preparation types — are included in our general assumption, we can conclude:

To the extent that a system C is given the ability to develop and time to vary, the vital series and its medial changes from any initial compositional value gain more and more importance toward *pure mediation*.

Second Chapter.

The Variation of the Final Changes.

I.

375. — For the other case to be considered now, and the further development of the final change of higher order, we first have to highlight its special requirements.

Let Rx be any environmental constituent to which any system C is exposed. We assume that any Rx-related property br is repeated in Rx at each setting — otherwise Rx may be the same or a similar environmental component. Under this assumption, br would be the most imaginable repetitive (recurrence) of Rx.

But we can also assume that any condition bc associated with the system C is repeated in every statement of C — that is, it is conceivably the most self-repeating (recurring) of the system C.

376. — So far as Rx is a complementary condition for all (primary) final properties which C will set when Rx is set, the condition br is also conceivably the most self-repeating (recurring) of this complementary condition. But if the system, as the epitome of the systematic preconditions, is also a condition for the final condition given by Rx, the condition bc is the conceivable self-repeating (recurrent) of the systematic preconditions, the totality of which represents C.

377. — Now C (as the epitome of the systematic preconditions) and Rx first form the conditional group (see no. 55) br and bc thus also make up what is conceivably the most repetitive (recurring) part of the conditional universe.

378. — The final form, which sets C when Rx is set, must therefore be presumed to be conditioned on the most self-repeating (recurring) of both classes (*Klassen*) (the environmental constituent Rx and the system C as the epitome of the systematic preconditions); but it is not immediately to be

assumed that this final form embraces all that is conceivably the most self-recurring classes both conditionally and exclusively.

379. — Not *complete;* the final form which encompasses the most conditional reoccurrences of both classes, would presuppose a very definite system change for which (according to no. 99) Rx could initially be claimed only as a conceivable complimentary condition. However, it does not follow from this that all of the components that make up br are included in every law of Rx, that each of these components is also a true complementary condition for that system change, ie., end state form would have to be addressed; For it cannot yet be concluded from the mere positing of a component that it is already set under such special conditions as to secure the meaning of a change condition for a given system C in all circumstances.

380. — Not *exclusively;* Just as the relatively largest preparation must decide on the composition, and also on the composition of each final condition (see no., 300), every composite end condition is determined by those forms of change which are most prepared at the time of settlement. So it could only be assumed pure, conditioned on the most conceivable repetitive end conditions, if no other conceivable forms of change for the system C had the value of most prepared which, in contrast to the other forms of change, are only conditioned by the most repetitive components of the associated conditions.

381. — But since the dependence on the conceivably most repetitive components of the combination of surroundings is not the only conceivable way of preparing the system C, then the more admixtures which are not conditioned by the conceivably most repetitive of both classes are to be presupposed: on the one hand other forms and forms of preparation are presumably involved in the determination of the final condition, and on the other hand, the number of practicing cases is assumed to be lower.

II.

382. — If we now call an end condition of the system C, which is not exclusively constituted by the conditional repeating of both classes as idiosyncratic, it can be assumed that the endings complementarily conditioned by a clearly determined combination Rx of space — as far as the system C itself comes into consideration — correspond to the concept of individual differences (*Idiosyndema*):

The more formal qualities, which do not belong to the most often imaginable ones, are congenital to the particular system C (be it as a result of inheritance of any individual peculiarities acquired by ancestors, or as a result of their own pre-birth acquisition);

the younger (that is, the less change conditions set by Rx, the more accessible the type of change condition at all) the system C; and the more the social circle to which the system is assigned (family, community, tribe, people, state, church), forms of changes which are not conditioned by what is most likely to be repetitive, and which has been preserved according to the intimacy of its own interconnectedness – is passed down further.

383. — It is in accordance with these conditions, that involvement of change forms, which are not conditioned by the most repetitive, are also assumed to be in such an end condition, coincides with a partial-systematic Komoment Γ; so, in turn, the preservation of the same in that (particular) end condition is to be presupposed the more, the more limited it is e.g., to presuppose:

On the one hand, the ability and the time of system C to develop in a positive direction (belonging to an ancestry that is no longer or no longer capable of development or to a degenerating ancestry; individual inability to develop due to lack of equipment or standstill that has occurred, degeneration initiated, suppression of the facility due to unfavorable living conditions, early demise, etc.), on the other hand, the conditions of extension and further development of what is conceivably the most repetitive, and the limitation and backward development of what is not so conditioned.

384. — And vice versa, the elimination of the non-conditional from the conceivable, mostly repetitive, from an end condition of the value Γ must be presupposed — the more so, the less limited — e.g., the less avoidable of system C's ability and time to positive development in general, and the conditions of limitation and reverse development of the conditional forms of change, as well as conceding the conditions of the extension and further development of such conditions.

III.

385. — Now, however, there are the conditions for the extension and further development of the most often-recurring conditional components of an idiosyncracy, e.g., for the system C limitation and regression of the not so conditional it is presupposed:

A) in the multiplication of the components dependent on an environmental combination in general by progressive exercise; specifically

1) in the increasing formal and functional differentiation of the system C (see nos. 73 and 118);

2) in the increasing approximation of every conceivable end-condition to the existing initial condition of system C by the repetition of its conditions (see no. 288);

B) in the proliferation of the dependent vital series approach to more and more immutable set endings in general; especially:

1) in the positive co-moments of self repeating dependent components (see no. 258ß)

2) in the increasing negative co-momentation of the co-moment variations dependent on an avoidable combination of components (see no. 257)

3) in the increasing positive co-momentation of the co-moment variations, which are dependent on an unavoidable combination of components (compare no. 256).

386. — Provided that each vital series develops an end condition Γ, which was developed within a certain combination circle (compare no. 312), it either eliminates the deviations as variations of the moments or raises irremovable deviations to new moments (with the original variation as the added integrating component) ($\Gamma + \Delta\Gamma$ positive moments), which approximates the development of the values of a subconstant (see no. 325). Thus, any combination of the environment into a single subconstant belonging at any given time, and thus the individual vital series itself, tends to form and preserve the components which depend on the uniform recurring component.

387. — If, then by further practice both vital series and its individual members (by analogy with nos. 372, 384) tend to be limited to the indespensible; by creating (relatively) new (differentiated) multiponibles with more restricted combination circles, but everywhere the deviations become more expendable; Thus, the subconstants developed by System C with respect to the same environmental combinations in the course of the further exercise, form once again a series, e.g., a subconstant series: thus the (individual) subconstant series is even more decisive than the single vital series for the formation and preservation of the of uniform self-recurring conditional components.

388. — If we pass from the individual human being as an individual of lower order to human societies as entities of higher order, which not only survive individuals but also partly take over and continue their achievements, we obtain developmental series of a higher dimension than their members individual subconstant series. These subconstant higher-order series will tend all the more toward the formation and preservation of uniform repetitive components, the more viable and stable the congregational system is -, at the same time, the broader it is, so that the finite nature differences, which are based not on environmental differences but on peculiarities of individuals of the lowest order, appear as opposite deviations at the place of their meeting the individual systems $C_1, C_2, \ldots, C_n$, with co-moments becoming more and more negative.

389. — Since, according to this, the multiplication of the spatial and temporal conditions of development of the final qualities also increases the elimination of the idiosyndema e.g., individual differences, (*Idiosyndema*) which presupposes the components of unconditioned components that are presumed to be the most repetitive of both classes; It follows that the finality of the value Γ, which realizes a system C when a clearly defined combination of environments is realized, will be the more so as to be presumed to be conceivable on the repetitive conditional: in a later time the individual is born and is added to the family into which it is born, the next circle of society expands to humanity and the next to the environment of the whole of the continents.

390. — Thus, the more the conditions for positive system development of individuals, generations, peoples, and humanity in general, the more they are assumed to extend to each individual system over time and hardly expand, the more the end conditions of the value Γ approximate the pure, conceivable and mostly self-repeating conditions.

391. — If we take the expression "function" in the mathematical sense, the latter result can be expressed as follows:

The approximation of the end qualities from the pure value Γ by the conceivable repetitive conditions is to be assumed as a function of space and time.

Third Chapter.

The Pure Constants of the System C.

I.

392. — According to the theorem (see no. 188), that the system C, if at all conceived to be fully assertive within certain limits, must be thought of changing until the vital difference is abolished, is followed by endings which enter the higher vital series order and can be introduced by virtue of their

preparation, without canceling the vital difference of a higher order (see no. 224), nor can the vital series be completed — that is, in this sense (as final the change) it is untenable.

393. — If it is assumed in a certain case, e.g., if a recurrence of the environment combination Rx, which had already been set several times, was given as a condition of the higher-order vital order, then this set of vital series for a particular individual living in a linguistic community in a certain time completed with the final condition $\Gamma\omega$; but it follows from this assumption that at that for the individual at that time this $\Gamma\omega$ canceled out the vital difference of a higher order.

394. — However, it does not follow from this assumption that the $\Gamma\omega$ setting exists for the same individual at other times, nor that $\Gamma\omega$ exists for any or all individuals with who live in a linguistic community, or at any time or even at all times that the value of the end state in relation to a higher-order series of nerves initiated by Rx; Thus, whenever Rx is given as the condition of a higher-order vital series, $\Gamma\omega$ can also be set invariably as the final end condition, having the greatest ability to be set. (*grössten Setzbarkeit*)

395. — An end condition of the value Γ is now conceivable for the greatest possible number of cases of the setting of Rx and the Rx-conditioned vital series (higher order) of the largest possible number of individuals connected by linguistic communication and for the longest period of time, as the final end-state invariably settable, that is, in relation to R, is conceived of the greatest possible settability, would be called a *perfect constant* of the system C. (in the same sense as no. 373)

II.

396. —Suppose that *pure constant* with respect to Rx, let it be Γk; thus Γk should have the following properties:

1) Since (according to n. 57) the form of that end condition of system C, that is, the formal value of C + ΔC, must depend on C on the one hand as the quintessence of its formal and functional initial structures, on the other

hand, from the formal nature of the condition of change, as a result of which it is increased by ΔC, Γk — insofar as its associated final condition form must be thought of as a partial systematic comoment at the same time — must not have absorbed partial forms which are based on individual systems; in fact:

a) neither on an individual difference of the system C_1, which is increased by ΔC in a given time; nor

b) another system C, whose changes to its own system could become a further alteration condition by virtue of the linguistic connection.

397. — If an end condition Γk should be thought of as a *pure constant* for Rx, then all the form differences of Γk must be determined by individual deviations of the systems C_1, C_2,. , , Cn could be conditioned on each other and eliminated, and the determination of the form Γk by the system C must be assumed to be limited only to the dependence on its most often repetitive nature bc (see no. 375).

398. — 2) Γk may not engage in any partial forms which would not find their formal condition in any conceivable setting of Rx; This means

a) always, when the given Rx is set, those R-values whose conditions are the partial forms of Γk must be included - as far as they are conditioned on R-values;

b) and, in any case, they must not be set differently - in terms of quality and quantity - than they are set in the specific cases.

399. —Therefore:

If an end condition Γk is to be thought of as a *pure constant* for Rx, then all differences in form of Γk, which could be caused by individual deviation of the individual cases of the Rx laws, must be eliminated, and Γk as a whole must be composed only of such partial forms, which are due to the associated individual case components of the most conceivable repetitive nature br.

400. — But even in its conceivable condition, being mostly repetitive, and consisting of the corresponding combination of conditions Rx, an end

condition can only be ascribed to the greatest possible immutability of its composability, and at the same time corresponds to the most self-repetitive, most complete, most accurate and simplest composition.

401. — The above can be summarized and generalized in the following sentence:

If an end condition of the value Γ is to be thought of as a *perfect constant* for a clearly defined combination of environments, then it must be thought of only as most conceivable, mostly repetitive of both conditional classes, but also in the most exact and simplest way possible.

402. — Now (according to no. 391) the approximation of the final qualities from the value Γ to the pure is to be presupposed by the conceivable self-repetitive condition as a function of space and time; However, the condition for the greatest possible settability of an end condition of the value Γ is that it is purely conditioned (exclusively) by the conceivable, usually self-recurring, condition.

403. — But since the activity - in accordance with the spatial and temporal increases in practicing R-values — the completeness, precision, and (by eliminating the expendable) the simplification of the final textures (see no.385) — also increases.

404. — Or (according to no. 395) shorter:

The approximation of the end condition from the value Γ to pure constants must be assumed as a function of space and time.

III.

405. — In the result of our last analysis we have only a development of an earlier one, according to which the recurrence of the conditions of settlement caused the recurrence of the settability (the multiponeity) (see no. 310). The greatest possible settability is equivalent to the (conceivably)

greatest temporal and spatial unlimitedness and inevitability of setting an end condition of value Γ as a conditional: and this is a function of space and time; meaning (in the words of no. 388) it is conditioned by the greatest possible temporal and spatial infinity and inevitability as its conditions of settlement.

406. — If, we finally call a vital series, which is provided with a *perfect median change* (no. 373) as a medial change and with a *perfect constant* as a final change, and a perfect series, then the overall result of this section is the sentence:

The development of system C varies the vital series in the sense of approaching a *complete series*.

Eighth Section

The Variation of the Independent Multiples of the Highest Order through the Further Development of the System C.

First Chapter

The Approach of the Multiponibles of the Highest Order Possible to a Pure Constant.

I.

407. — We now turn to the special case of an end condition in general, which as a conditional complementary condition cannot be set by the setting of this or that surrounding component, but is set by any element of the environment, ie., it is conceivable as a multiponible of the highest order.

For brevity we refer to this multiponible of the highest order with the symbol Γy.

408. — The more it can be assumed that a particular system C has developed under *individual* circumstances; the more the settings of the multiponibles Γy belong to that system at a given time, which developed under other circumstances of origin and social circle e.g., unavoidable.

Now let us consider the limitations and avoidability of the settability of historically developed individual multiponibles $\Gamma y\,(1)$, $\Gamma y(2)$. . . $\Gamma y(n)$ as diminishing, and their quality must also be approximated to that of an independent perfect constant.

II.

409. — On the assumption that the conceivable repetition of both the entire environmental components and the entire systems C_1, as long as humans are

accepted in linguistic communities on earth, it can likewise be assumed, that the conceivable approximation of the multiponibles Γy to a *perfect constant* can also be assumed for our purposes, e.g., to greatly simplify their determination.

If, for example, we assume that the condition of conceivably mostly repetitive qualities in the systems C and in the surrounding environment is shared, then we must consider both properties as immutable, because their positive or negative multiplication would contradict the assumption.

410. — According to this supposition, the conceivably most repetitive of system C where it is presupposed in system C that the most self-repeating end of the environment, provided they are in any case an alteration condition for system C, is also presupposed as a condition of change for C; Thus, whenever and wherever system R (see no. 42) and system C are presumed together, a form of change of C is necessary, which depends on the conceivable repetition of both classes and presupposes the largest possible environment — namely system R itself — and the greatest possible arrangement.

411. — But also consider the greatest immutability imaginable; because it finds the conditions of its settlement always and everywhere absolutely uniform, e.g., no (positive or negative) increase of the temporal or spatial conditions of change of any kind could increase the forms of exercised expression (positive or negative).

412. — If, however, this form of change of system C, conditioned in this way, can be ascribed to the greatest possible temporal and spatial unlimitedness and inevitability of its ability to be set, then the exclusiveness of its positing cannot immediately and indefinitely be presupposed; rather, it must first be assumed that they (see no. 378) will be only a partial form (partial determination) of a more complex form of change, which system C realizes within the development of (hu)mankind: and that, therefore, even the final condition associated with it is only a constant component of all multiponibles of the type Γy, and will not merely constitute this multiplicity of itself alone.

413. — This means that the historically developed individual multiponibles Γy (1), Γy (2), Γy (n) ... are to be presupposed as individual differences (*Idiosyndemes*) (see no. 382), which are the most repetitive of the two classes containing ending conditions, among others, which are not conditioned by the mostly self-recurring ends of both classes.

III.

414. — On the other hand, we denote this constant component of all multiponibles Γy as a, and those components which are to be presupposed as not conditioned on the two conceivable mostly self-relieving conditions, with α, and finally that determination of the multiponibles Γy, which depends on both kinds of components, e.g., its idiosyndemic composition, is denoted with y; Thus the analytic expression for the historically developed individual multiponible of the highest order conceivable is:

$y == f(a, α).$

This formula would be sufficient, if it concerns only the analytic expression for the individual Multiponible's conceivable highest order. As regards the more precise formula, which would have to take into account the specific qualities of what is conceivable as the most repetitive, non-conditioned, and the corresponding cases of the determination of these subjects as idiosyndemes, see Drobisch's presentation of the "common law" of a "series of similar subjects", describing the "successive qualities of one and the same changing subject") (see note no. 8).

415. — But since (because of our assumption) the variability of a is excluded, any approximation of the multiponibles Γy to a *perfect constant* can only be thought of as a diminution of α.

416. — And since finally the reduction of the limitation and avoidance of the settability of historically developed multiponibles Γy(1), Γy(2), Γy(n) ..., is conditioned by the approximation to the perfect (pure) constant, this approximation is conditioned by the diminution from α, it follows:

If the limitation and avoidance of the settability of historically developed multiponibles Γy (1), Γy (2), Γy (n) ..., are to be reduced, their value α must be reduced.

IV.

417. — All conceivable multiples Γy(1), Γy(2), Γy(n) ..., which differ by the a-values, can be ordered according to their decreasing a-values into a first series, their first term thus being determined by the maximum value of α, the last of which is formed by the minimum value of α.

418. — Further, we can think of all the multiponibles Γy(1), Γy(2), Γy(n) ..., already historically assumed in the course of human evolution, as different forms of change of respective set values according to their chronological order in a second series.

419. — And finally we can think of a third series, if we put together from the second (series) those members which coincide with members of the first series, in the same order prescribed by the first series.

420. — Now (according to no. 415) the approximation to the *pure constant* can only be thought of as diminution of a, but the diminution of a itself determines the direction of the first series; There is thus, a further change of historically given multiponibles Γy of the kind that can only be thought of as approximation to the *perfect constant*, if it coincides with the direction of the first series.

421. — Since the first and third series have not the number of members, but the direction in common, it follows that any historically realized or yet to be effected alteration of historically presupposed multiponibles of the highest order Γ(x)y — and thus every increase of the second series which does not at the same time continue the third series — does not approximate the *perfect constants* in the historically assumed multiponible I(x)y.

Second Chapter.

Application of the Vital Series of Higher Order.

I.

422. — Of all conceivable cases, which would likely result in an increase of the series of historically developed multiponibles $\Gamma y(1)$, $\Gamma y(2)$, $\Gamma y(n) \ldots$, only those who would introduce a (relatively) new multiponible of this kind due to the setting of a higher-order vital difference and with the function of annulment it would be considered for our investigation. — So, we have yet to apply the last sentences to the special case that the multiponibles of the type Γy belong to any vital series and are bound to their further development.

For we can make the assumption that in a special case the end condition, which belongs to the initial change of a vital series (higher order), is at the same time a multiponible of the type Γy; then posit the further assumption that the system C completely asserts itself under the reduction of its set vital conservation value.

We can express these assumptions simply: Let system C set the task of completing a vital series (higher order), by a multiponible of the same kind., in which a multiponible of the type Γy is introduced.

423. — Firstly, this general task can be made that system C only ever reach an end state which completes the vital series (higher order), and which also possesses the meaning of a multiponible Γy for the individual system C, without any eventual increase and diminution of its establishment: a task of the first order.

But one can also narrow the task to the point that the final condition associated with the final change, which the vital series can individually conclude, is at the same time a multiponible of diminished limitation and inevitability of its establishment: a task of the second-order.

And finally, one can sharpen the task in such a way that the final condition in question is of the greatest possible scope and inevitability of its establishment (*Setzbarkeit*): a task of the third order.

424. — But since every series of vitalities arising from a historically developed multiponition Γy (x) presupposes a change in them, we must, under our assumptions, consider either a or a changed.

But not a, since (according to no. 414) this was assumed to be unchangeable; consequently a.

II.

424. — But since every series of vitalities arising from a historically developed multiponition Γy (x) presupposes a change in them, we must, under our assumptions, consider either a or α changed.

But not a, since (according to no. 414) this was assumed to be unchangeable; consequently α.

425. — It follows:

1) The solution of the task of the first order is conceivable only as the positing of a multiponible Γy which at the same time abolishes the notional change of a;

2) the solution of the second-order problem is conceivable only as the positing of a multiponible Γy which at the same time abolishes the notional change of a by the simultaneous diminution of a;

3) the solution of the third-order problem is conceivable only as the positing of a multiponible Γy, which at the same time removes the notational and the conceivable changes of a by repealing a.

426. — For the solution of the task of the first order any number of manifold endings are conceivable, since any change which (according to no. 299) in sense and scope of system C can be considered as the establishment of the

final change, insofar as only the final condition associated with it, in its form, at the time of its setting, corresponds to the formal conditions of abolition of the vital difference for the individual.

427. — The conditions for solving the problem of second order are somewhat more complicated, since it is not only a question of achieving a reduction of the deviation (of the partial systematic co-moment) to zero, but also of eliminating any α-values; but here, too, System C still has a wide scope of individual possibilities.

428. — On the other hand, the task of the third order only allows one solution. The reduction of the deviation (from the partial systematic co-moment) to the value zero must go together with the reduction of the value α to zero; but only a multiponible Γy is conceivable whose value α had the value zero: namely a multiponible Γy purely from the value a. Since, however, a multiponible Γy is the constant a purely by value, or in other words the conceivably purest multiponible Γy is set so, the task of the third order can only be solved by setting a final change purely from the form of the constant a.

429. — Since, finally, *approximation* to the *perfect constant* (according to no. 415) is only conceivable as a *reduction* of α, but the reduction of α reaches its extreme limit with the minimum of a, e.g., with a value of $\alpha == 0$, meaning, with the pure constant a; So any approximation to the *perfect constant* is an approximation to the pure constant a. In other words, in the case that the multiponible Γy should be thought of as a *perfect constant* at the same time, the pure constant a is to be thought of as the *perfect constant* for the multiponible Γy.

III.

430. — Since the final condition is the value of a, and not its isolated setting (according to no. 412), everything that causes the reduction of α to zero must also prepare the isolated positing of a; meaning to realize the conceivable approximation of the multiponible Γy to the perfect constant a.

431. — However, the general conditions for the reduction from α to zero are all the more presupposed in the negative co-moments of the less uniformly practiced forms and the special conditions:

on the one hand the more mutually exclusive and thus negatively committing α-values are supplied to the same system C;

and on the other hand, the more special cases of environmental constituents, provided that their placement initiates a vital series, *complete constants (vollkommene Konstanten)* also emerge which give off special and particularly well-functioning ("effective") final changes for the possible vital series, so that the final functions of individually determined final qualities are dispensable and thus functionally eliminated.

Both special conditions for our special case are fulfilled by the development of *perfect constants* in general (no. 404).

432. — And this means that:

The more the original limited environment of human individuals on earth, and the different human individuals of different order to a human individual of the highest order imaginable, namely to a total time of positive human, development in general, extended — the more the general development in the positive direction approximating the multiponible Γy of the *pure constant a* and the human individual, for which it annuls any positing of the vital series derived from the individual (*Idiosyndem*) $y == f(a, \alpha)$ to mankind.

We denote the content of no. 432 as the theorem of progressive elimination.

433. — If one did not want to allow for the assumption of the constant a in the historically developed multiponibles $\Gamma y_{(1)}$, $\Gamma y_{(2)}$, $\Gamma y_{(n)}$ ..., then only the simplification would be lost in the conditions of approximation to the *perfect constant* — the simplification, namely, that the self-repeating condition of both classes is already established everywhere in the special cases of multiponables Γy, and therefore did not have to fulfill the conditions — yet to be paid — of production. But since these latter conditions are altogether contained in the general conditions of the approximation to *perfect constants* (see no. 385), the result remains the same:

A multiponible Γy, which should at the same time be thought of as a *perfect constant*, is to be thought of as conditioned by the conceivable, and usually repetitive, of all the surrounding components and of all the systems C and the approximation of historically developed multiponibles $\Gamma y(1)$, $\Gamma y(2)$ ... $\Gamma y(n)$ ... this *perfect constant* exists as a function of space and time.

Since the inevitability of this development follows from the inevitability of its internal conditions, all the moments which determine the direction of the second series (no. 418) from that of the first (no. 417), or more generally: the direction of further activity is refracted from the direction of the *perfect constant*, as they are considered from a higher point of view mere *external* conditions of change and describe the deviations as mere *developmental disorders*.

---

434. — Examining the multiponibles of the highest order possible may end the analysis of our environment and the changes of its most important components, namely the conditions of change of the genus R and the variable systems of system C.

We have undertaken this analysis in order to find an answer to the question: in what sense and to what extent can components of our environment be accepted as prerequisites of experience (no. 40).

Since the experience, as it is put in the question, would be dependent on those changes to which certain constituents of the environment are a precondition (see no. 1), so the general result for us is that if any R-value is accepted as a condition of change for a human individual and thus for system C, then an experience is stated under the E values dependent on C + ΔC, that R-value can be related to the experience in no other condition than C + ΔC (see no. 91).

435. — This means:

Components R_1, R_2..., R_n will be thought of as a prerequisites of predicated experience, thus they can be accepted as such only in the sense of complementary conditions for the final properties of the system C, and indeed only insofar as statements of content (E-values) can be thought of as dependent on these final qualities; in this case, for all final condition determinations (*Endbescheffenheitsbestimmungen*).

Attachment
"Criticism of Pure Experience" by Dr. Richard Avenarius

The Graphic Presentation of the Fluctuation of System C.

Excerpted from the Quarterly Journal for Economic Philosophy, XVII, 1893, p. 500

——

The representative curve of the fluctuating course of System C is provided for didactic purposes; I published it after the wishes of a few scientific friends; it serves to illustrate the usage of the process.

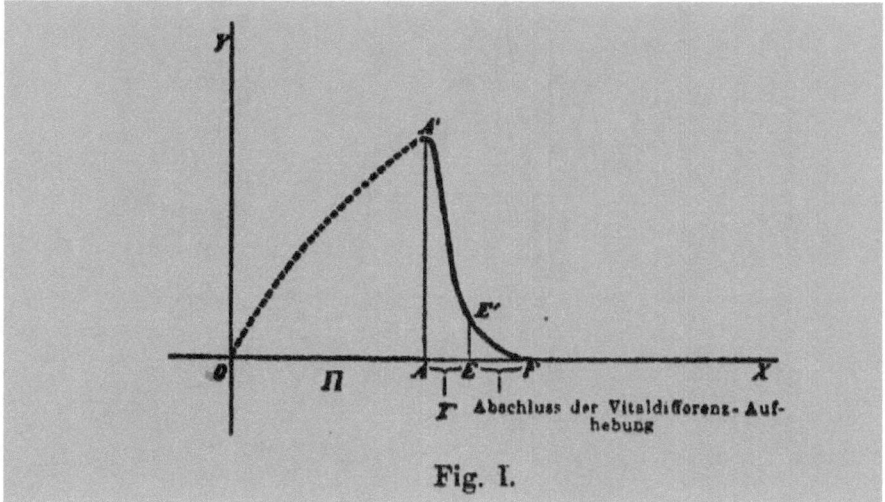

Fig. I.

see "Abschluss der Vitaldifferenz-Aufhebung" trans: Conclusion of the vital difference nullification.

Fig. I Illustrates the case that assumes nutritional increase of value Π the main partial of System C occurs with the associated complement at the end condition e.g., the final completed form contributes to the partial moment Π standing in relation to the partial moment Γ. (Critique of Pure Experience, sec. 201).

Under this condition, Γ represents that increase in labor which with Π defines the vital difference of the first order (sec. 203) to the annulment (*Aufhebung*) which slows at the conclusion of the process (in the selected case as well as instances where there is conceivable acceleration to the end point.). Let us now bear on the abscissa axis of a right-angled coordinate system YOX the time, on the ordinate axis at each point in time corresponding to vital differences, so it is at O the partial System C, to think about the vital conservation maximum. The nutritional increase Π set during sleep is shown by the ascending branch of the curve; it is (in accordance with the execution of the schema in Criticism of Pure

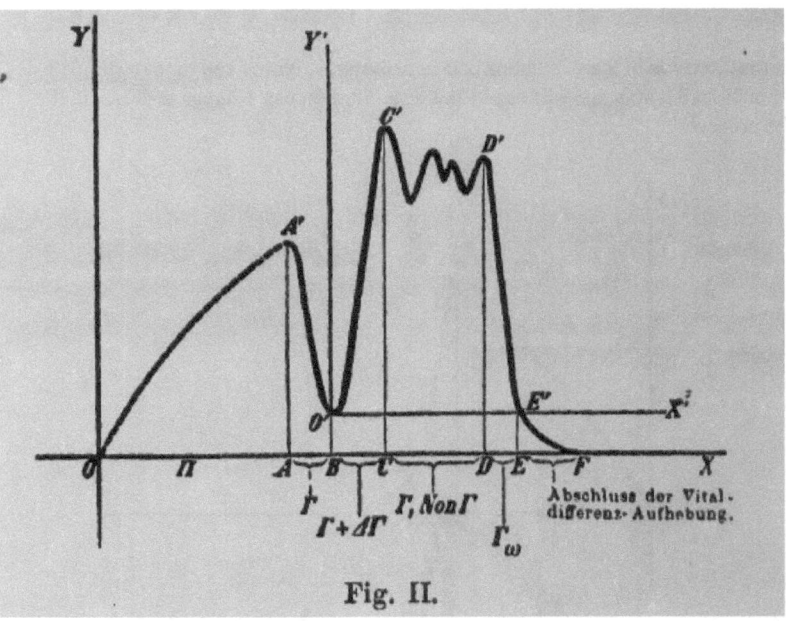

Fig. II.

Experience II. p. 438, n. 502), to indicate that no statement content is associated with it. The curve now rises up to point A, which corresponds to the time point A, in which the complimentary condition is realized the end condition, Γ; then a relative steep drop until point E yielding the distance A->E which represents the approximation of System C to the annulment of the vital difference. According to the slowdown assumed for the present case, the curve runs from E until it reaches the abscissa axis at F; the route E->F corresponds to the time part in which the annulment of the vital difference has become perfectly complete (compare num. 182). On the axis of the abscissa the final route is set O->F the first order vital series; the path O->A is the initial section, the path A->E the medial and E->F the final section.

The case represented by Figure II deviates from the assumptions made in the first case (Figure 1) the condition of completeness sets the variation of (point) Γ. The analytical expression of this variation is Γ + ΔΓ. Alone it is further assumed that (in accordance with num. 326) that the trained form Γ reaches a time portion before ΔΓ will be achieved. Thusly we obtain the composition Γ, Γ + ΔΓ; and that means an explicit vital series introduction (num. 220, 323, 326). The higher order vitality difference is set herein (num. 204) and is created by the annulment of variation Γ (num. 227) the latter annulment after a more or less notable change in the included values with other end values (non- Γ) is achieved (num. 326). The end condition itself, in which the vital difference of higher order is cancelled, takes the place of the variable Γ and thus the vital series of higher order ends - the final unit (n. 182, 326) - is designated Γw.

The course of the curve from point O -> A is the same as in the previous case; the distance A-> B on the abscissa axis shows the time in which Γ was set before Γ + ΔΓ. Only at the point B is the variation of Γ itself is realized; accordingly, the curve increases from O again, reaching the assumed time point C - the highest point. The timepoints from C-> D illustrate the alternate approaches and distances in the partials System C in relation to its vital preserving maximum. The slope of the curve from D-> E represents the annulment of the vital difference of the higher order, shown at the time of D-> E, while in the E-> F time frame the vital differential annulment is completed; the course of the curve itself from E-> F is again the same as Figure I.

It is therefore in the flow of first case (Fig. I) that the oscillation switched into a second variation (higher order) which begins at O and ends at E.

The latter curve builds within the YOX coordinate system on the originating ordered curve; the second coordinate system YOX delineates the curve of the switched on fluctuation, namely OCDE against the originating curve shown in Fig.1. The case in Figure II now consists of two separate pieces OAO and EF. The distance AE on the abscissa axis finally represents the vital series of higher order according to their temporary successive sections; namely the route AB pre-section. The individual section endings result in the higher order vital series according to their membership: namely the series Γ; Γ + ΔΓ; Γ, Non- Γ; Γw (see table num. 327 included below) (e.g. first order, higher order).

Number 327 – "Kritik den reinen Ehrfahrung" Erste Teil (VOL1) by Richard Avenarius

Number 327 — "Critic of Pure Experience by Dr. Richard Avenarius (First Part) VOL I translated by David Grunwald

Case 1: Vital Order	Initial change	Middle change				Final change
Case 2: Higher Vital Order		Sub-section	Initial change	Middle change	Final change	
Overall vital difference	1st Order	1st Order	Higher order	Change	1st Order	0 Order
Members of the series	Partial system movement Π	Γ1	Γ1+ ΔΓ1	Γ1 Non- Γ1	Γw= Γe od Γw Γo	Annulment of fluctuation
Fluctuation size	The system O only enters into the conditions of being awake after the	Sufficient for significance	Sufficient for significance	Sufficient for significance	Sufficient for significance	

	fluctuation has been set					
Fluctuation form	...	As a multiple, it depends on the repetitive components of the environment	Reduction of the multiplicity with respect to the previous associated environment combination	Change	Approach to a sub-constant	_____
Fluctuation relevance	...	The substantial fluctuation belongs to a main partial system	The substantial fluctuation belongs to a main partial system	The substantial fluctuation belongs to a main partial system	The substantial fluctuation belongs to a main partial system	_____
Fluctuation direction	...	Fully negative	Changes in positive direction	Change	Cancellation of directional change	_____
Fluctuation variation after _____ - form	...	Partial system fully trained component	Positive trans-exertion	Change	Negative trans-exertion	_____
Fluctuation variation after _____ - value	...	Maximum proficiency	Decreased proficiency	Change	Increased proficiency	_____
Fluctuation variation after _____ - context	...	Minimum articulation e.g. opposition	Maximum articulation e.g., opposition	Change	Decreasing of articulation e.g., opposition	_____

Annotation

1. A series of related cases if included in W. Grieinger in his "Pathology of Psychiatric Illnesses" 8, Braunschweig 1871, p. 88 a reproduced case is noted here Esquirol reports. In 1816 there was a 38 year-old Jewess in the Pitié-Salpêtrière afflicted with mania and blindness. Despite this she saw the strangest things. She died suddenly; I found the *nervi optici* fully atrophied.

2. (p. 39, n. 78.) Physiology uses the expressions "sensitive", "sensual" and "sensory" without sufficient distinction; in the text we differentiate the meaning in of sense for our task.

3. (no. 95) Since these are not "psycho-physical" measurements, we disregard any so-called "reaction time" for the sake of simplicity.

4. (number 118) See W. Wündt, Grundzüge der physiologischen Psychologie 8 , Band I, Leipzig 1887, S. 224 f., 241 f. and especially page 331.

5. (no. 158) The presuppositions, which are united in the theorem of number 158, are scientific common property. Suffice it to remember here: not only that overhaul changes the central partial systems pathologically, the same goes for the the diet meaning in their preservation, when they are out of action, that is, they are short of work - - A very general observation, which is even used for the experimental determination of the central course of the nerve tracts ((GUDDEN method.) On the other hand, the anatomical findings are chiefly hyperaemia, that is, an excess of nutrition. And finally, for example, as obstruction of cerebral vessels - so too little nutrition - on the preservation of the associated area.

6. see number 262, compare. Wündt, see above vol II , p. 408.

7. (no. 327.) The preceding propositions about the vital series put to the reader perhaps present *foreign* demands, the changes of man, by which he moves within a non-ideal environment (see no. 187), without thinking about the further assumption of a "consciousness", and it would perhaps be my task, to explain the independent vital series itself, which is not already well *known* in the same sense and measure, such as the preparatory changes, the influence of the lack of work or the lack of nutrition on the nervous

structures and the like. But before this could happen, I would have to draw the reader's attention to what must be expected of him as well, and what alone he should expect from me.

In the latter regard, above all, let him not *demand* proof that certain changes in an individual are really "unconscious." This "proof" is as impossible as would be the opposite; but it is also completely unnecessary: because it can only be about how we can *think* about these changes. The restraint requested here of the reader is thus only a methodological one; It can be granted completely without prejudice to the systematic question: whether in truth a "consciousness" should be accepted at the same time or not. However, as we have learned, the "growth" of plant and animal organisms, their existence and growth, their coloring, their nutrition (in this respect it may incidentally be remembered in particular that the animal organism is within certain dietary limits and that in warm-blooded animals the nervous system mediates an increase in the metabolic rate, depending on the lowering of the external temperature [see L. Hermann, Lehrbuch der Physiologie 8, Berlin 1886, pp. 204 ff.]) — I say we have learned to be able to think specifically about the nutrition of organisms, their healing after injury, their recovery from illness, their adaptation to environmental changes, etc., without the "participation" of "one spirit" or "of the mind." It is also necessary to acquire the ability to be able to think about the so-called "appropriate" changes and changes in the system C, without immediately calling for a "spirit" to explain it, especially since "changes in mental state" are still yet to be explained.

The demand to think of the so-called "purposeful" movements of the limbs, the changes of the expression of the face, speaking, etc. purely from changes of a nervous central organ, will be difficult or even impossible to fulfill for those who are too one-sidedly used to thinking these movements are guided by a "mind" or a "consciousness".

For this reason, and for the first time, only those series of changes may be chosen for the explanation of the independent series of vital factors, the arrangement and course of which may be practiced without the intervention of a "consciousness." It will then not be difficult to apply the presented conditions to the human system C even if they do not directly affect the human being or even specifically the brain. Whether or not the changes to be made to that lower system of nervous features called spinal cord (including the medulla oblongata), but ultimately related to "sensory functions", is completely irrelevant to our purpose: these movements were done with "consciousness", yet this does not prove that they themselves came *from* "consciousness": that they were made *from* "consciousness," would be, I mean, in the sense of those physiologists who still believe in the "sensory functions of the spinal cord," but are well aware that the "changes of consciousness" on which the movements depend in any case, reflect changes in the material substratum: as one can occasionally reflect on the "changes of consciousness", with no regard for changes in the "*material substratum*", one must also be able to reflect on the changes of this so-called "material substratum" of eventual "psychic concomitants."

For those psychologists then, who are indeed trained to grasp the "manifestations of consciousness" according to *mechanical principles*, should take heed, to understand even more complex "functional" series of changes of system C in the same way, for these psychologists I notice that it must be completely irrelevant to the "effectiveness" of *mechanical principles*, whether the variable is a "nervous form element" or a "conscious idea", since it cannot depend on the variable, but only on the changes in the application of the mechanical approach.

The series of changes I present here for illustration, are well-known - at least to the physiologists; after all, they should be quoted here in detail, so that they are really available. I borrow the cases especially noted, Ed. Pfleger, who gathered facts in his interesting Monograph: "The sensory functions of the spinal cord of the vertebrates "(Berlin 1853); since then the experiments have become part of common physiological lecture series. — Regarding the author's thesis, whether it approves or disapproves me, I have no reason to comment here. Only in general it can be remarked: who in his thinking about the "consciousness" of the brain may as well find it "arbitrary", not too "attributed" to the spinal cord, that if they "accidentally" had had the "non-consciousness" of the spinal cord as the fixed starting-point of their thinking, they would only be able to do so by an act of "arbitrariness," all at once to "subordinate" the "brain" "consciousness".

Also, every logical — demarcation with "consciousness" as part of the central nervous system inside the spinal column and then transformed into the upper vertebra of the cranium will always appear somewhat "arbitrary" against those areas containing "consciousness" and vice versa to those areas without "consciousness."

Now to the illustration itself!

I. a) Not infrequently, the frog, if only the spine is separated, but the head is not completely removed, " sets its paw against the the upper part of the neck wound with one of its hind feet, as if it had wanted to break its head completely away from its body"; Indian pigs, soon after beheading, "place the hind paw on the neck wound and rub against it." Also kittens rub their neck wounds soon after decapitation. If one pinches the paw of the beheaded frog, it pulls it back; if you repeat it, it hides the paw under its stomach and withdraws as if in fear, into itself. If one prods the animal more intensively with knives and tweezers, it grabs them with its paw or pushes the objects back altogether."(p. 15., p. 26)

The frog made repeated attempts, when the cloaca was irritated, to remove the instrument with the hind legs. "Once," explained Pflüger, "a small board was placed on the foot of the decapitated frog and an attempt was made to attach a string for an immobilization experiment, but the animal continued to brush it away with the other hind paw, and despite its relentless efforts to strip away the whole apparatus, I was eventually successful."

Decapitated turtles when similarly irritated became stuck in their shells.

When the frog, pinched at a certain point with tweezers reacted with gripping and stemming movements, while the irritation of the same spot by a droplet of vinegar gave a different order of action: the hind leg moved to the irritated skin area and rubbed on it back and forth.

Fr. Goltz (Contributions to the teaching of the functions of Nerve centers of the frog. Berlin 1869, p. 116) performed the experiment with the following modifications: he broke both thighs of the decapitated frogs and brought them to the prone position and in each case irritated the skin close to the midline by brushing on acetic acid. In almost all cases the broken limbs reached for the acid etched spot. The same results were achieved with a decapitated frog with an artificially shortened thigh.

A beheaded toad, held over a burning candle turned back and forth and took hold of the tweezers, which were held by the forefoot "in order to get away from the flame." If the tails of eels and earth salamanders are cut below the medulla oblongata and placed in burning fire; even the tail itself turns away from the fire. The same was observed in young suitably prepared kittens; one sees "clearly and definitely, that the tail avoids the fire from the right, bending the tail to the left and if pursued further to the left side of the body, the tail is pulled as cats usually do."

If the thigh is stretched away from the beheaded frog, the animal almost always pulls him back to the body; the beheaded salamander sooner or later puts one foot back to its normal position from where it was placed; "Place the hind leg a little back so the *dorsum pedis* touches the ground and the animal will not maintain this position but bring the foot back to normal." Lay a decapitated earth salamander and eel on its back and it will place itself back on its stomach; under the same conditions the same occurs for pieces of animals consisting of just two legs and a tail. And more! The part of the animal seeks to attain balance, "you can see how the legs part, if irritated again to see whether it would lie down on its back again and how it moves as if in painful moments striving to maintain its balance.

Male frogs (Rana temporaria when cut while mating holds the female spinal cord tightly between the atlas and second vertebra and does not let go of the female. If one tries to move his arm, "he only embraces her tighter and places his arms deep inside her breast." If some vinegar acid is placed on his arm, the frog uses the other arm to hold the female and cleans the corrosive substance with its hind legs. "Afterward he holds the female with both arms."

b) If the medulla oblongata has been left to the terrestrial salamanders, they will automatically rise again, placed on their backs; Frogs, which only have the cerebellum, the elongated medulla and the spinal cord, turn back in the same way from the prone position and do this even under aggravating circumstances, such as if a rear limb was sewn to the trunk of the animal. (Pflüger, Goltz)

W. Wundt mentions on the occasion of his report (a. Bd. II, p. 495) on the movements of partially revealed animals, that birds whose brain lobes have been removed, even then - apparently spontaneously – they cleaned their feathers, he adds: "it is hard to doubt that such movements activated by skin irritants also cause the same movements in uncut animals."

Especially interesting is the experiments seen for the first time performed by Goltz: the frog's cerebrum is removed and it is placed on a board or on a flat hand and tilted in such a way so as to simulate the animal falling. The frog "then bends his head forward all the way thereby approaching the center of gravity to support the itself. If the surface is rotated it crawls up the inclined plane, comes to the edge of the board or hand and sits. If the board or hand was turned completely downwards, the frog moves to the surface area, either upwards or downwards (to the back of the hand).

The brainless frog thus "maintains itself with great skill" even in in such predicaments the balance is firm." And he needs "for executing the balancing act" not only vision, but a center of equilibrium at the four points in order to be able to work successfully. This "requires sense of touch imparted at the skin. The ability to assert weight therefore is immediately lost if for example the animal's skin is removed from the hind limbs." — Incidentally, if the movement of the support surface is so hasty that the animal must make a greater effort to balance, it jumps off.

Similarly, pigeons without a cerebrum have the same balance behavior. After a report by Rosenthal a pigeon was "set on the edge of a long table on a line. As soon as the bird's foot was set into empty air, it began to beat its wings until it's both its feet were again on the table. The bird was made to walk to the other side of the table to renew the same experiment. This went on for an hour or longer with the same regularity. I carefully placed the pigeon on a horizontally held finger, where it remained sitting clasping the finger with its talons as birds do on poles and twigs. But as soon as I turned my finger around its axis so that the animal's head was tilted, the dove began to beat its wings, thus protecting itself from falling, always adjusted itself to the twisted finger ... "

II. a) In the cases cited, we have relatively simple series of changes whose initial element is a peripheral change in the normal behavior of the central nervous system; the end result is the annulment of the initial condition /element (cleaning the acid, pushing away the tweezers, backing away from fire, restoring to a familiar position (e.g., equilibrium); these movements form the middle stages. Such change series can be called relatively simple, because the middle condition completes in a relatively simple manner. The following cases may show more complex series in which the lower nervous system — the maintenance of the functionality — moves from more experienced to less experienced and from simpler to less simple changes until such time has passed that the initial change is cancelled.

A frog has been decapitated below the medulla oblongata and a skin close to the condyle of the internal femoris is irritated by the application of a droplet of acetic acid. "The result is that the frog bends the irritated leg, the other stretches, so that the body is slightly pulled over to the stretched legs. The foot of the irritated thigh is guided with the dorsum of the toes against the irritated parts of the skin, with this dorsum, he wipes off the corrosive substance as the foot is abducted and adducted. In order to see what would happen when "the old movement no longer clears the substance, Pflüger cut off the lower leg of another similarly treated frog. At first the frog applied the same remedy: "if the small area of the skin over the condyle of the internal femoris is irritated, the irritated thigh is hunched, the unruly leg stretches and the stump of the lower leg moves in a way that makes it undoubtedly the same appearance here as it used to be ". These movements do not lead to the target while the stimulus remains — now the lower system attempts new and more complicated movements: the movements generally became "restless", "different" movements became "purposeless"; until finally a movement "quit often" achieved success. Sometimes this happens in such a way that the frog "bends the irritated thigh much more than before, because he still had the lower leg, so now, after the trunk is bent forward, the irritated thigh, which is also still rotated outwards, so the side surface of the torso can be wiped off." In other cases, the transition to different and "unfamiliar" movements is even more striking: the irritated leg (the amputated lower leg) is stretched, the unruly leg against is moderately bent and adducted - and finally with the sole of the adducted foot the corrosive acid of the irritated thigh is wiped off. "One sees," said Pflüger, "how these moves (the original) are completely different from the previous movements. In the previous movement, the flexion of the irritated and non-irritated extension was present; in this case just the opposite occurs, namely the extension of the irritated and flexion of the non-irritated leg, although only one skin spot had been irritated in both movements."

Another illustration offers the following experiment, which Pflüger himself describes as a modification of the above. — A beheaded frog is placed on its stomach and acid is poured along the back on the left or right side. If the irritation comes from the right, it uses the toes of the right root to wipe off the acid. If the left side is irritated, the same is done on the left. Now a leg of the headless frog is cut off — let's assume it's the right one. Now, if acid is applied on the right part of the dorsal skin, the left leg wipes off the acid on the right backside. (p. 124)

b) An experiment of the same kind as that which Pflüger employed on beheaded frogs, and on which we have just given a lecture, was also carried out on a sleeping three-year-old boy with the same success: on the tickling of the right nostril, the child raised his right hand in a defensive movement, then rubbed his right nostril; when tickling the left nostril, the left hand is used. Now Pflüger quietly placed both arms of the child sleeping on his back next to the body and cautiously prevented the left arm from being guided to the face. Now the left nostril of the youth was tickled again: the left arm immediately moved again, but could not reach the irritated spot. The boy grimaced, and then, as the irritant remained brought up the right hand to the left nostril.

c) The following variation of the experiment on the frog is reported by Auerbach and reproduced by Goltz (Auerbach in Günsburg's Journal of Clinical Medicine, Volume IV, Book IV, p. 487; Goltz p. 111 f. - See also Wundt p. 490).

The beheaded frog has been laid on its back and a sheathing has been placed over a calf or the plantar side of the tarsus: the animal stretches out both legs and moves them at the same time against each other and rub their lower parts with the Plantar side together by alternating bending and stretching movements. The animal struggled hard to get rid of it, but I held its foot and dabbed the above-mentioned spot with sulfuric acid ... In the first haste, indeed, the animal missed the irritated spot; very soon, however, it straightened the leg more at the same time it led far to the left side, reaching the aching area, which was then rubbed in this unusual position."

The next two attempts are again by Goltz himself and may be narrated in his own words. "I fastened the trunk of a beheaded frog on a board in the prone position. The arms were fixed immovably on the surface. Then I sewed on the skin the Achilles tendon of the right leg together with the Achilles tendon of the left foot. If I fasten the left foot somewhere on the board, then the right leg is also fixed. Depending on the location where I fix the left foot, so the right hip and knee joint correspond. I fasted the left leg so firm that the right leg could not move either. The axis of the right thigh was 90 degrees from the torso. The thighs and lower legs were so close to the knee joint at an angle of 70 degrees. The animal was immobilized in this position with only the right foot movable at the ankle. Now I brushed some acid on the right side of the midline. If I stimulate the unbounded limb, the leg becomes strongly bent at the hip joint and knee joint and the dorsiflexion at the ankle easily extends to the agitated area. In this case the hip and knee joints are unfavorably placed in a fixed angular position. I told myself that the animal only carries out the movement on the freed ankle, otherwise it could never reach the agitated spot. But what happened? Since the dorsiflexion is not quite enough to reach the area, the animal hyperextends its toes. The foot which is fixed is bent so strongly that its dorsal surface shows significant recess. As a result of the fixation, the upper joints show an unusual adjustment in all foot and toe joints. In such way, the tips of the toes reach the irritated spot and perform imperfect movements there. . .

"I placed the beheaded frog in the prone position as in the previous experiments, but leave both legs free (unattached). I place the thighs of the animal on each side so their axis coincides with the midline of the trunk and forms an angle of around 110 degrees. Close to the skin of the popliteal fossa on each side, I insert a cylindrical fastener which protrudes vertically. The fastener is not allowed to harm the animal's skin. Then I bring both knee joints into an acute-angled flexion, so that every protruding fastener is placed across the hollow of the knee. The animal remains calm in this situation. Acetic acid is applied to the skin of the outer ankles and the outer edges of both feet. The movement with which the unrestrained beheaded frog responds is often described. The animal draws both feet together behind the trunk midline and rubs the feet against each other. Also in our case here, we see the same movement initiated; but the ordinary approach in this case cannot be accomplished as the fasteners do not allow the leg a straight path to the rear of the midline. Although futile, the twitching movements continue for awhile and the legs are always pushed against the fasteners. The frog's legs work against the fasteners as if trying to push them away. Suddenly the animal performs a movement quite different from the previous one. The animal pushes the knees to the front of the torso, thus removing them from the fasteners and now moves the feet between the fasteners without inhibition."

d) GOLTZ also commented on frogs that had their cerebrum removed e.g., a book was used to jab the right thigh. "After a few defensive movements and repeated jabs, the animal jumps or crawls on and avoids it by taking care of the obstacle by avoiding it on the left." Goltz now brought the animal back to its original position, put the obstacle in the way the frog had just chosen and irritated the animal "again same body part, and behold, he now goes a whole different way. It goes straight then crawls around the obstacle on the right side. At times Goltz even saw "the animal, instead of bypassing the obstacle sideways, was able to make a jump over the low book. Goltz now took the frog without cerebrum and attached the right posterior arm to the body which made it unusable." The frog had the same success: circumventing the use of one of its main limbs, the frog avoided the object in its path. It knew, despite the disruptive interference in the musculature apparatus, that with the remnants of the forces left over it achieved the specific goal of avoiding an obstacle.

To examine expected changes Goltz severed the optical nerves and carefully pushed the air out of the lungs of the frogs without a cerebrum. The animals rose from the bottom of a water filled container as a result of shortness of breath after a longer time than a blinded frog that still possessed a cerebrum, but nevertheless "quite in the same way". Goltz then made modifications to the experiment, transitioning to a more complicated experiment which he describes as follows: "In a large vessel filled with water, I put a bottle filled with water in the opposite direction, so that the water in the bottle was held by the pressure of the atmosphere. I place a blinded frog into it without air. The animal rises in the bottle and touches the bottom of the bottle with its nose. After some time the animal responds to the need for breath it gropes restlessly on the walls of the bottle and finally finds the large mouth of the bottle to escape from it. I now tested with one brainless frog in similar predicament. The case was quite similar. This one also found its way out of the labyrinth."

III. Details of these experiments deserve mentioning:

a) Page 205, line 21 ff. Of our report, we have a case in which a decapitated animal uses a circumstance just presented as a means.

b) Page 206, line 11 ff., We find another case in which a means that already serves a "purpose" is recorded.

c) In a third case, the remedy is taken after the experimenter does the exercise in the particular direction (only once). Pflüger reports (following the P. 207 below): "The lower leg is amputated and when a droplet of acetic acid brought to the definite place close to the condyle of the internal femoris, it is seen sometimes as restless, searching movements of the animal, that it will not find the right solution. If one then grasps the foot of the unrendered leg pushes it against the irritated thighs, but without touching the site the wetted with acetic acid, the frog will, if you let go of it, itself take the way suggested and lead the foot against the irritated place and wipe it off. "

d) In a fourth case, about which ship (according to Auerbach) reports (textbook of human physiology, I. Lahr

1858-59, p. 218), a remedy developed on a later occasion will be applied retroactively to an earlier "irritation": After beheading and amputation of a thigh dabbed with acetic acid on the corresponding backside. "The animal, deprived of use of the corresponding thigh, becomes very restless and finally, tires, then remains silent. If you later dab a spot on the other half of the body, the animal wipes it with its foot, and after that happens suddenly he attacks the other side with the same foot and rubs the spot first affected.

e) An influence of both positive and negative practice is also stated by WUNDT (a, a, vol. II, p. 492). In positive regards: "The amputated frog, having once had the leg on the other hand used to remove the corrosive substances, makes it *easier* to do the same thing again in future cases. A certain obvious practice can occur. In a negative sense: "If in the experiments in which obstacles are intentionally opposed to the execution of a certain movement, a longer time elapses between the action of the stimuli, one sees again and again the same fruitless efforts of the finally successful correct movement, and In many cases, this does not come about. Here the mechanical facilities have been lost."

In passing even an "isolated spinal cord piece is still able to learn" is shown by SCHIFF with the attempts of Flourens at Tritonen, whom he the "Halsmark" had intersected (provided that a re-association really did not occur): the animals moved "the hind legs after several weeks and even more after months much more regularly and more moderately than at the beginning ".

f) An influence of the speed of stimulation illustrated – after WUNDT - the following experiment carried out by Goltz (Wundt, p. 493 note; Goltz p. 127 ff.).

There are two frogs, one beheaded and the other blinded. They are placed in a vessel that is filled with water so only a small part of the animals protrude; then the water is heated gradually. Both frogs sit quietly until 25° C; From that point, the blinded frog becomes uncomfortable — its breathing quickens, the movements become more agitated, moving from surface to immersion, making desperate leaps until — the water temperature climbs to around 42° C — the breathing becomes intermittent, there are tetanic convulsions and the animal dies.   Meantime the headless frog sits without any movement, when its back is agitated with a little acetic acid its rear feet make the appropriate movements and return to their former place. The animal sits motionless and with the heat the muscles bend the animal forward — "the headless . . . is without any signs of life and died." WUNDT adds the remark to his paper: "This experiment shows very clearly how the mechanism of the spinal cord react to such stimuli according to the general law of nerve excitation, which act with a certain speed, while a gradually growing stimulus remains completely ineffective."

g) A case where the repeal of an original change condition and a new condition is to be found among Goltz's experiments with the "reflexive activity of the frog's vocal condition. Goltz had made the discovery that frogs whose cerebrum was severed would emit a croaking noise when excited. He reports that "if you apply constant pressure on the back of a cerebrum deprived frog, the animal croaks for awhile until it calms down. "If you remove the pressure the frog croaks unusually after having been silent for a while. The removal of the stimuli appears to have acted like a new stimulus."

h) On page 205 lines 5, we have a case of "dependency on the form of irritant." Goltz makes the following remark on the occasion (see p. 4): "As stated you can facilitate the croaking only through peculiar mechanical agitation and pressure or by stroking the skin. Of course, you can use any smooth rounded objects instead of the finger to arouse the skin. But not every form of mechanical arousal is suitable to cause the croaking. If I use the tip of the tool to scratch the back, the animal makes defensive movements, but does not croak. The croaking is also not triggered by chemical agitation. I apply dilute acetic on the back of the animal, and familiar wiping movements ensue, but the animal makes no sound. Even electrical application on the dorsal skin proves ineffective. The animal flings the electrodes away without screaming. So only a very specific form of mechanical irritation triggers croaking, namely applied pressure or stroking of the back skin with a smooth surfaced object.

i) If what is said concerns the "stimulus form" by the same author, the following is on the influence of "stimulus intensity" in the brainless frog: "Usually, touching the animal anywhere lightly makes no movement. If the contact is more forceful, pinched or jabbed, the animal follows known defensive movements. If the stimulus is made even stronger, you can see movement in the whole body which is different that when the stimulus was less forceful. If the stimulus is very strong or it is repeated often, the animal moves to·jump from it.
The law of movements can be clearly seen depending on the strength of the stimulus. If one touches at the cornea of the eye of the cerebrum deprived frog with a needle, the first and next movement, the animal responds by closing the eyelid. Repeat the same irritation in succession whereby one does not look at whether the needle meets the eye or the eyelid and the animal pushes the needle away with the forefoot on the same side. This is the second form of movement the animal responds with. With continued and more intensive mistreatment of the eye, the animal turns its head and the upper part of the trunk after the opposite over, and finally, when the stimulus increased and strengthened, the animal moves away from the place. The order of these four motion sequences were not always maintained, but in some animals the regularity in the modification was notable. "For example, one such frog always took the needle away with its forepaw when the eye had been touched three times. The first two times it simply winked."
B. Luchsinger also conducted experiments on the influence of stimulus on forms of beheaded and headed frog's and the results of Pflüger's archive for the entire physiology has been published, Bonn 1880, Bd. XXII u. XXIII. The author explains reflexes generally (which he explains about spinal cords, e.g.,

innately acquired through activity): "the crossed reflexes are not specific to a stimuli, mild stimuli of any kind may trigger them; with stronger stimuli, however, also modest reflexes arise and these can blur the crossed ones." Then in particular (Vol. XXIII, p. 309 ff.): If you behead a Triton and hang it on a tripod floating freely in the air, you can see mechanical stimulation of the tail and if the jabs it with the tip of a knife it gives a familiar movement. In the same way, but less easily - this succeeds when attempted on a newt or lizard or eel; so it is with a snake. A snake is beheaded. Its tail is stroked gently which causes it to turn; jabbed and the tail turns away more so ... In his well-known experiment, Pflüger examines the movement of eel tails under normal conditions; On the other hand, TIEGEL only perceived the turning of the tails in snakes. Our attempts to reconcile the differences between the two researchers leads to attribute the results to differences in the intensity of the application of stimuli. But the normal spinal cord knows how to respond to these different stimuli in different but always expedient ways. Weak stimulus is approximate for the tails the same for intact animals. Strong stimuli are countered by averting the tail, an escape symptom of a healthy animal. The taxation of stimulus strength is dealt with differently in different animals. A strong stimulus for one animal may be a weak one for another one. . . A frog with the cerebrum or even mid-cerebrum removed can be laid on his back and return to the prone position. If the animal is placed on his back he returns immediately; placed on a smooth surface and let go of gradually, the frog can stay in an abnormal position for a long period of time. Only additional stimulus can move it. This change can happen in two ways depending on the force and direction of the applied stimuli. In fact, if you gently rub any part of the trunk, the animal turns toward the stimulus and rotation occurs when the abdominal surface turns in such a to the stimulus. But if we approach the brainless animal with a burning match, the reversal takes place in the opposite direction and the abdominal surface turns away from the stimulus. Accordingly, a different goal is required for each different muscle action. When turning towards (the stimulus), the hind leg of the other side is essential, but when turning away, the hind leg of the same side is involved. So here too we see a different behavior of the alter nervous system against external stimuli. Analogous behavior was shown by other animals like salamanders and newts.

The animal too, knows how to adapt itself perfectly to external circumstances.

IV. As shown in the cases it follows that nervous systems are highly organized, and we can think of the higher series changes with well-determined initial, middle and end members (cite cases under 1a and 1b). Our specific methodology demanded highly organized nervous systems for setting higher order changes — and those cited were such nervous systems: namely those without "consciousness" and "consciousness" then proceeding from physiology.   Those readers are already well versed in thinking of certain pathological conditions as "unconscious". Incidentally, there are whole groups of cases in which higher order change series appear set: the "expedient actions" during epileptic seizures, in certain hypnotic states and in cases of brain lesions.   I close with the oft-quoted words of Spinoza: *"Hitherto it has not been determined what the body can do, that is, I have been taught no experience of the natural laws of the body, in things corporeal, of what can be and what cannot be, unless it is determined from the mind."*

8. number 414 M. W. DROBISCH, Neue Darstellung der Logik 4 Leipzig 1875. S. 190, § 151.

TRANSLATOR NOTES

Notes to Volume I of "Critique of Pure Experience" ("*Kritik der reinen Ehrfahrung*").

Volume I is a complete detailing of Dr. Richard Avenarius' system C, and the objects which form the foundation of the system — a complete system of knowledge.

Capturing Avenarius' meaning while maintaining his loose "exploratory" style was challenging to say the least. Perhaps this is one of the reasons it has never been successfully done in its entirety. In his forward he makes it clear he himself is attempting something new. In his open, exploratory language, he experimented with philology - adding suffixes and creating new words. His use of technical grammar like semicolons and em dashes helped in clarifying his meaning but raised challenges to grammatical purists. It may be added that the criticism Avenarius anticipated will also be levelled to his first and only translator.

A background in linguistics, humanities, literature and computer science aided the translation as I came to believe in the power of Avenarius' system and the strength of his humanist/scientific vision. As far as I am concerned, this book should be required reading in the 21st century for poets and scientists alike. In the end, Dr. Avenarius found an understanding of himself. Isn't that the greatest gift of all?

The original linguistical forms often used forms (dafs, dass) that required careful editing and were rendered as close to the original as possible. The ideas and description belong to Dr. Avenarius, the mistakes of course are mine.

The "Einleitung" to volume one was composed of 10 sections containing 39 numbered subsections. The first part (Volume 1) consists of section 40-435 (395 in total) followed by Annotations and a graphical presentation of the fluctuations of System C completed with a graph and charts capturing the main concepts. This will complete the first volume. The second volume will be presented as a separate book.

In composing his work, Avenarius took great care to utilize earlier sections in elucidating ideas and strengthening points as he went. The constellation of ideas then forms a complete system of knowledge. Volume one consists of eight sections which head various chapters. The end of the first volume provides an annotation and a graphic presentation of the fluctuation of System C. Inside each chapter, Avenaruis includes notes, often distinguished by lower sized fonts. Other methods of adding stress and elucidation include: use of bold, italics and foreign languages which have been preserved as best as possible from the original printings of the third and final edition released in 1921.

In the English translation key or unique German terms are included in italicized parenthesis with reference numbers to the text or glossary. The complete bibliography can be found in the companion original in German for each volume.

Avenarius' use of references predates hyperlinks. He utilizes references to numbered segments and sections demarcated by roman numerals. For purposes of clarity, references to numbered sections will be recorded as (num 1,2,3,...) and for sections (sec 1,2,3,...). Another decision was made to allow the multiple phrases Avenarius uses and for two reasons. The first is to preserve the "voice" of the explorer. The reader will read Avenarius as he thought. Secondly, the often laymen's simplicity his ideas were spread out makes them accessible and like system C itself, able to instruct as sub-ideas as well as in the context of a whole. The careful reader will gain an appreciation of a first-rate mind and hopefully gain new insights in the world in which we live as well as a person's role in it as they struggle to survive.

D. G. Campbell, CA 2018

www.ingramcontent.com/pod-product-compliance
Lightning Source LLC
Chambersburg PA
CBHW030937180526
45163CB00002B/603